IMPROVING PRODUCTIVITY OF TUBE MANUFACTURE IN NON-FERROUS METALS

IMPROVING PRODUCTIVITY OF TUBE MANUFACTURE IN NON-FERROUS METALS

Yu F SHEVAKIN AND A M RYTIKOV

Translated by Major J H Dixon, F.I.L.(Russ.)

Published in Russian by "Metallurgiya" Press, Moscow 1968

METINFORM

TECHNICOPY LIMITED

STONEHOUSE GLOS ENGLAND

Printed and published by Technicopy Limited, Stonehouse Glos.

PREFACE

The directives for the five-year plan approved by the 23rd Congress of the Communist Party of the Soviet Union (CPSU) envisage a considerable increase in labour productivity. For instance, by 1970 the production of steel tubes (seamless and welded) will be 50-70% higher than tube production in 1965. The increase in the production of non-ferrous metal tubes will be equally great.

Non-ferrous metal tubes (cylindrical and special shaped) are extensively used in different types of power plant. This improves the efficiency of this type of plant, increases its power per unit mass of equipment, and increases its life, while at the same time saving metal. Much use is at present being made of various methods of producing non-ferrous metal tubes. One of the most important tasks, therefore, is that of rationalizing both the process and the technology combining different processes.

After analysing new experimental data and producing a great number of batches of tubes for balance purposes, the authors were able to consider some of the problems involved in extrusion, cold rolling and drawing in greater detail than in their earlier work ("Production of ferrous metal tubes" Metallurgizdat 1963). The more precise relations between the different parameters of the processes made it possible to work out a procedure for calculating rational technologies and to assess different methods and schedules from an economic point of view, also to decide upon methods of raising output.

The laws which were found can be utilized for developing draft designs for new tube production shops, also for assessing the present technologies, and principally for developing new technologies for use at tube works.

The book considers the production of tubes made of heavy non-ferrous metals and alloys. The data given can, however, also be applied to the production of tubes made of steel or light metals and alloys.

CONTENTS

INTRODUCTION

Tubes are at present produced by a variety of methods; of these the methods most used are hot and cold rolling, extrusion, drawing and welding. In most cases the technologies for producing tubes include several operations, methods of pressure treatment, carried out in succession. The selection of the technology and method of treatment is dictated by the physical properties of the metal concerned, the required quality and dimensions of the product and in many cases the tonnage of the batch, the availability of appropriate equipment, and its rational use.

The following are the processes most used at Soviet works for producing tubes made of heavy non-ferrous metals: extrusion, cold rolling and drawing. The principal advantage of the extrusion process is the stressed state, consisting of non-uniform compression from all sides (1-4); this makes it possible to raise the plastic properties of the metal and to carry out further processing with considerable degrees of deformation. Extrusion is used for producing cylindrical tubes, also hollow sections (5,6), which cannot be produced by other methods.

When the metal has been processed it has a fine-grained structure and high mechanical properties. The cost of the tools required is comparatively low, and they can be changed comparatively rapidly; these factors make the extrusion process versatile and particularly effective in the production of comparatively small tonnage batches of products (4,7).

The types of press at present most used in the production of tubes made of heavy non-ferrous metals are vertical hydraulic presses of 6-15MN (600-1500 tons) and horizontal hydraulic presses exerting a maximum force of 10-80MN (1000-8000 tons) (5). Presses with higher maximum forces of 120-150MN (12-14000 tons) or over are being used in the production of light non-ferrous metal tubes. According to reference (7), the use of 200MN (20000 tons) presses will make them competitive with cogging and blooming mills for the production of steel semi-products. (All tons referred to in this text are metric tons).

In most cases the capital outlay on building extrusion shops is lower than the outlay on rolling mills. They take up less space. Extrusion is used for producing either in tube cold rolling mills (KhPT) or on draw benches. Presses of two standard dimensions are sometimes used for producing tube blanks, for instance: horizontal presses 25-30MN (2.5-3000 tons) or vertical presses 6MN (600 tons).

The cold rolling of tubes is a comparatively new method (the first KhPT mills were constructed in the USSR in 1937); they achieved recognition because of certain advantages which they possess. The principal advantage of the method is that tubes can be rolled with high reductions per cycle by comparison with drawing (the reduction is μ = 6-12 or more). This shortens tube production cycle times, and eliminates a number of technological operations. The cold rolling method of producing tubes is competitive with drawing, particularly in the production of tubes made with metals and alloys with poor ductility (9).

Modern methods of designing the grooves in the rolls of KhPT mills (8) make it possible to roll tubes with quite close limits as to both external diameter and wall thickness. The initial relative variation in the thickness of the walls of tube blanks during the rolling process is greatly (50-67%) reduced (10,11), and this makes it possible to produce tubes with little eccentricity of external diameter relative to internal, and with high surface finishes (categories 8-10). Cold rolling has the further advantage that tubes with their cross-section varying along their length, and with different transverse sections, can be produced (6,12).

The most extensive range of tubes are produced by drawing (14-16). The high output of the method, the simplicity of the equipment, its low cost and the high quality of the tubes produced by this method make it an all-purpose method which is extensively used. Linear chain draw benches are the type most used. Drawing with bull blocks is also much employed. The high speeds achieved in drawing with bull blocks (up to 500-600m/min (18-19)) and the use of long tubes (up to 2000m) give this method excellent prospects owing to its high output figures and the fact that it is very economic.

In modern shops, tubes of one standard dimension can be produced by technologies utilising different methods of production. The wide range of present technologies can effectively be illustrated if we take condenser tubes made of L68 brass alloy as the example. These tubes are now made by four methods:

- Method I - double extrusion followed by drawing to the final dimensions. During the first operation of extrusion in a 2500-3000 ton horizontal hydraulic press the cylindrical tube blank, the "hollow" is produced. In the next operation the "hollow" is cut into measured lengths blanks for the 600 ton vertical hydraulic presses. A hole may be drilled in the blank in order that there shall be less fluctuation in wall thickness when the tube blank is extruded, or if the press is not fitted with an independent piercing system. The more modern technology is that of extruding a round blank with an aperture in 2500-3000 ton presses (21). "Bottle" mandrels are used for this. In this case the operation of drilling the hole is unnecessary. When the tube blank extruded in the vertical press has been pickled it is transported for drawing down to its final dimensions;

- Method II - extrusion of the tube blank in a 1500 ton press is followed by drawing. In this technology, by comparison with method I one operation is omitted (extrusion), and the operations of machining the blanks involved in this are also not carried out;

- Method III - the extrusion of tube blanks in a 3000 ton press followed by two cold rolling operations. The extruded blank is first prepared (pickled, straightened etc.) then rolled in a KhPT-55 or KhPT-75 mill to intermediate dimensions. When the tubes have been cut into measured lengths, annealed and pickled, they pass to a KhPT-32 mill, where they are rolled down to their final dimensions. It has been established that, in many cases, to raise the output of the KhPT-32 mills it is advisable to roll large diameter tubes with a wall thickness corresponding to the final dimension. In this case, as a rule a single drawing operation, usually without a mandrel, is the final operation;

- Method IV - the extrusion of the tube blanks in a 1500 ton press followed by rolling in a KhPT-32 (or KhPT-55) mill down to the final dimensions, or alternatively by rolling a semi-finished rolled product following by sizing drawing. By comparison with method III, this eliminates the operation of preparatory rolling in a large KhPT mill.

To make tube production more efficient, therefore, it is necessary to solve the following problems:

1) that of selecting the rational technological parameters for the operation;

2) that of selecting the optimum technology.

When either the first or the second problem is solved, the initial factors are the particulars of the equipment, the types available, and the range and volume of production.

These problems can most effectively be solved when a new shop is being designed and the equipment is selected on the basis of a prescribed range and volume of production. When the rational technological method and the parameters of a particular operation are being worked out for existing shops, the actual conditions (the equipment available etc.) must be taken into account. When rational technologies are being worked out, therefore, it is extremely important to establish general laws, valid in every case.

The selection of the rational technological parameters of an operation is based on a thorough knowledge of the process and its basic laws. For instance rational billet dimensions can only be selected on the basis of calculations of the piercing and extrusion forces, the rate of travel of the metal, and the temperature changes in the tube and billet metal, and by comparing all these quantities with the permissible values. When the parameters of the operations of cold rolling and drawing are decided upon, it is important to calculate the optimum dimensions of the blank. These calculations cannot be made without a knowledge of the drawing or cold rolling forces and without determining the outputs of the operations.

To establish the optimum technology it is necessary to take into account the economic aspects of production. The effectivenesses of different methods can be gauged by analysing and comparing the cost of production and its components for the different methods.

On the basis of what has been set forth above, the book has been divided into two parts: basic parameters of the processes of extrusion, drawing and cold rolling, and their calculation; rational technologies for producing tubes. In writing the book, the authors have attempted to give the analytical relationships obtained in a form such that they can easily be used in conjunction with computers.

PART I - MAIN PARAMETERS OF THE PROCESSES OF EXTRUSION, DRAWING AND COLD REDUCTION AND THEIR CALCULATION

CHAPTER I - EXTRUSION

The principal operations associated with the reduction of the metal are the piercing of the billet and the actual extrusion - the flow of the metal into the circular gap formed by the mandrel nose and the aperture of the die. The piercing of the billet preceeds extrusion, and centres the mandrel nose relative to the billet.

When a billet is pierced a proportion of the metal displaced by the mandrel nose flows out into the die aperture to form the piercing scalp, the dimensions (length) of which depend on the pressure head or the extrusion force, the relationship between the diameters of the mandrel nose and die aperture, and the other parameters of the process.

In extrusion, not all the billet metal is extruded to form the tube. That part of the metal which is not reduced is called the "head". The extent of this depends on the contact friction, the extent to which the metal cools, etc.

When press equipment is being selected and the technological parameters of the process decided upon, it is necessary to know the laws for the process, the relationships and interconnections between the different parameters, their effect on the quality of the product, etc.

BILLET PIERCING

During extrusion and piercing, the mandrels are subjected to extremely severe conditions. During piercing they undergo considerable longitudinal compressive stresses, and during extrusion they undergo longitudinal tensile stresses and radial compressive stresses. In some cases stresses caused by longitudinal bending of the mandrel nose are superimposed on these stresses.

In spite of the short time for which mandrel noses are in contact with the metal being extruded, after each extrusion cycle the nose has been heated to 300-700°C. The dimensions of tubes to be extruded, also the diameter and length of the billet, are in many cases determined on the basis of the force acting on the mandrel nose. This is most characteristic of tubes with internal diameters less than 30mm. The metals used for the mandrel noses, their heat treatment and their mechanical properties were given in references (5,16).

Piercing methods

Both Soviet and foreign scientists have published works on the process of piercing. Of these scientists the most important have been S I Gubkin, Sh Gelei, P S Istomin, M V Sotorozhev, E P Unksov, I L Perlin, I Ya Tarnovskii, V V Zholobov et alia.

There are the following methods of piercing:

1) open piercing (the most widely encountered variation of the first stage of three-dimensional extrusion);

2) piercing during the production of tubes by the Erhardt method;

3) piercing billets in a blind ended die (enclosed or incremental piercing), used in the production of tubes with an external diameter equal to that of the container;

4) piercing billets using a die with an aperture.

The last two methods of piercing are the most widely used in the production of non-ferrous metal tubes.

In many cases, in order to increase the production of sound metal when manufacturing large diameter tubes it is most effective (although the total extrusion cycle is slightly longer) to carry out extrusion and piercing using a "blind" die positioned before the main die in the container. When the billet is pierced the container opens, the mandrel nose pierces the billet (cuts off the bottom), and the "blind" die is forced out of the container. When the container has been closed by the die, the extrusion process proceeds in the normal manner.

Since the nature of flow by the metal governs the force conditions for the piercing process, the results of research and conclusions as to the forces acting on the piercing point must be compared very carefully for the different methods.

Deformation of the metal during piercing

Figure 1 shows two stages in metal flow during piercing. In the initial stage of the piercing process the metal mostly flows into the circular gap between the container and the mandrel nose, i.e. in a direction opposite to that in which the mandrel nose moves (fig.1a).

As the depth to which the metal is pierced increases further (fig. 1b), however, the amount of metal flowing into the die increases abruptly. This indicates a qualitative change in the nature of deformation. A volume of metal beneath the mandrel nose (the "plug") is separated from the remainder of the billet by shear deformation. After this moment the metal beneath the end of the mandrel nose is not deformed; this is confirmed by the small distortion of the co-ordinate grid in the "plugs". We can see from the illustration that the shear deformations are concentrated within a circumferential volume of metal with an integral diameter equal to that of the mandrel nose and an external diameter equal to that of the die aperture.

For purposes of comparison, fig. 2 shows how the co-ordinate

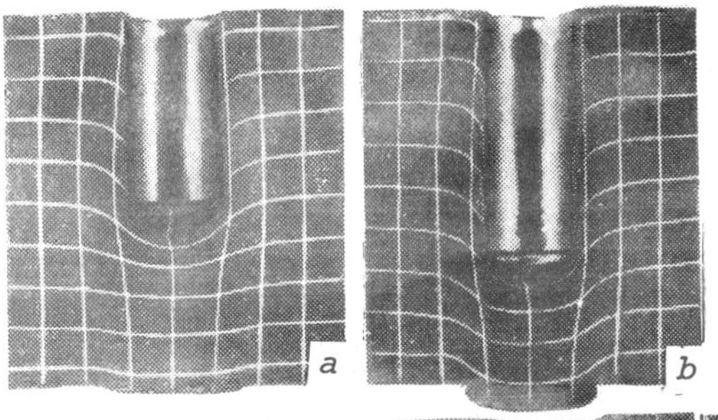

Fig.1
Deformation of the metal when a billet is pierced in a container with a die.

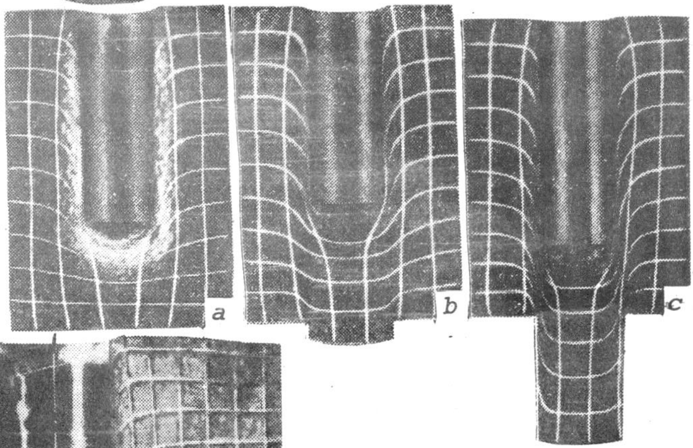

Fig.2
Distortion of the co-ordinate grid during piercing.

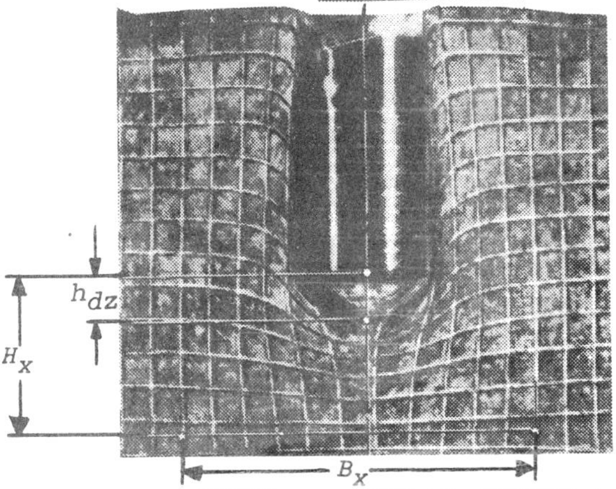

Fig.3
Parameters of the region of deformation while a billet is being pierced in a "blind" die.

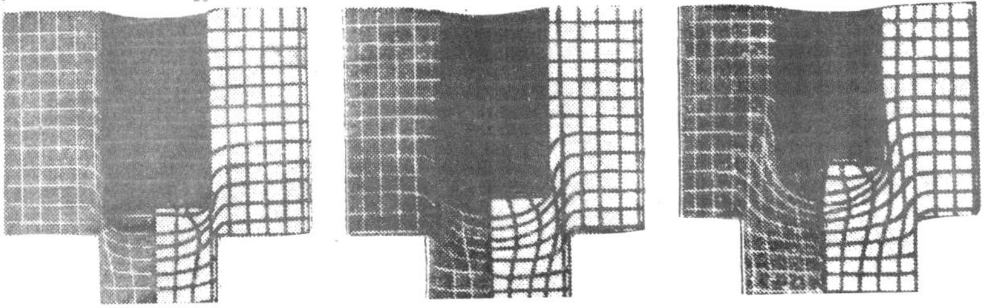

Fig.4 Mechanism of separation of the "scalp" in the deformation region with different die diameters. Container diameter D_K *= 41mm.*

grid is distorted when a billet is pierced (a) in a "blind" die, (b) in a die with an aperture, and (c) in a container enclosed at the top. Using the "blind" die the metal only flows in a direction opposite to that in which the mandrel nose moves; in the container enclosed at the top the metal all flows into the die. In the latter case shear deformation is present.

The piercing of a billet in a container and a die thus includes two characteristic successive stages of deformation: stage 1 is analogous with enclosed piercing, and must therefore be represented by all the laws for that process; stage 2 is the stage in which the metal shears and enters the die.

Research has proved that the dimensions of the "dead" zone at the tip of the mandrel nose in stage 1 of the process only alter slightly as the depth of piercing increases, and depend on the diameter of the mandrel nose. The relationship of the depth h_{mz} of the "dead" zone (fig.3) to the diameter of the mandrel nose is $\frac{h_{mz}}{d_{ig}} = 0.28\text{–}0.30$ (Table 1). As the diameter of the mandrel nose is increased, the extent of the zone of plastic deformation also increases.

As the depth of piercing increases, the ratio of the width B_x of deformation to the diameter d_{ig} of the mandrel nose does not alter greatly, and is in every case equal to about 3. The depth H_x of the region of deformation depends on the depth of piercing. This corresponds entirely to the results given in references (24,25).

Analysis of the distortion of the co-ordinate grid shows that, with $\frac{d_{ig}}{D_k} < 0.3$, the layers of metal adjacent to the surface of the container are not displaced relative to it, and that consequently contact friction between the surfaces of the metal and the container does not affect the piercing conditions. In this case the piercing process is analogous with the introduction of the mandrel nose into the slab with infinite dimensions. With $\frac{d_{ig}}{D_k} > 0.3$, the entire volume of the metal enters the annular gap between the mandrel nose and the container. In this case friction is bound to affect the piercing conditions.

The qualitative relationships for hot extrusion are greatly affected by the extent to which the metal cools in the contact layers. The ratios $\frac{d_{ig}}{D_k}$ found for hot extrusion may therefore be slightly different, but the qualitative picture will remain the same.

Fig. 5

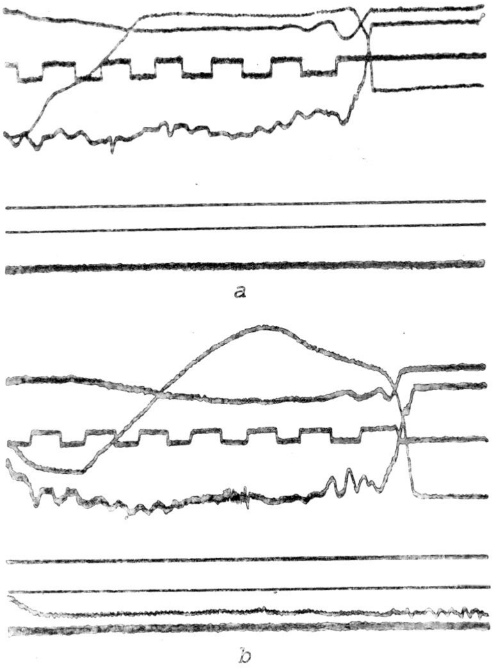

Oscillograms of the process of piercing billets with (a) cylindrical and (b) tapered mandrel noses.

Fig.6

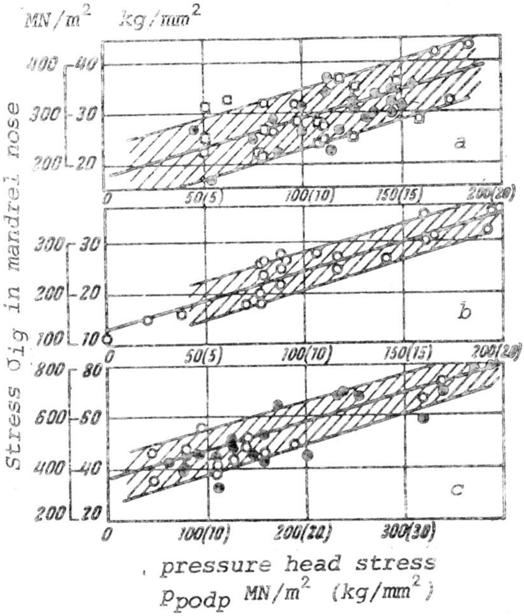

Relationship of the initial stress σ_{ig} in mandrel noses to the pressure head stress P_{podp}. 1500 tonne (15000kN) press: (a) copper billet temperature T_{sl}=800x850°C, container dia. D_k=180mm, tube dimensions: (○) 26x36mm; (●) 26x(33-34)mm; (□) 30x(37-39)mm; (b) L62 brass, T_{sl} = 730x750°C, D_k = 180mm, mandrel nose d_{ig} = 26-40mm; (c) MNZh5-1 bronze, T_{sl} = 900-920°C, D_k = 150mm, tube dimensions: (○) 30x38mm; (●) 35x(40-41)mm.

Table 1

Dimensions of the region of deformation in stage 1 of piercing (metal lead)

Process Parameters	Mandrel nose diameter d_{ig} mm											
	5			10			14		16	18		20
Container diameter D_k mm	36	41	41	36	41	41	41	61	41	41	61	41
Depth of piercing ℓ_x mm	14.1	14	18	14.8	15	21	15.3	14.2	19.5	14.5	14.4	20
Depth of dead zone h_{mz} mm	1.6	1.6	1.5	2.8	3	2.7	4.0	4.1	4.5	5	5	5.6
h_{mz}/d_{ig}	0.32	0.32	0.30	0.28	0.3	0.27	0.29	0.29	0.28	0.28	0.28	0.28
Depth H_x of region of deformation mm	10	11	7.8	12	12	11.5	15	16.1	16.5	16	17	18
H_x/d_{ig}	2	2.2	1.54	1.2	1.2	1.15	1.09	1.2	1.03	0.89	0.95	0.9
Width D_x of region of deformation mm	15.4	15.0	15.5	30.3	30.0	30.8	>41	42.3	>41	>41	55	>41
D_x/d_{ig}	3.08	3.00	3.1	3.03	3.0	3.08	> 2.9	3.02	> 2.5	> 2.28	3.05	> 2

Experiments (fig.4) proved that the volume of metal in which shear deformation takes place is close to that of a hollow truncated cone. The smaller the difference between the diameter of the mandrel nose and the aperture in the die, i.e. the smaller the tube wall thickness, the smaller will this volume be. If the shear deformation is regarded as shearing, the shear surface can be calculated using the equation:

$$S = \pi \frac{d_{ig} + D_{tr}}{2} \ell_{sr} \tag{1}$$

here ℓ_{sr} is the length of the region of shear (the distance of the mandrel nose tip from the die aperture).

Forces acting on the mandrel nose in piercing

As opposed to the process of enclosed piercing (24,27-29), the variation in the force exerted on the mandrel nose when a billet is pierced in a container with a die is represented by an extremes curve, the maximum on which is a certain distance from the die.

To make it easier for the metal to pass the mandrel nose during extrusion these mandrel noses are made with a slight taper. The tapers are given in Table 2.

Table 2

Taper of the piercer point $\Delta d = d_{max} - d_{min}$ over its working length l_{ig}

Press force tons	ℓ_{ig} mm	d_{ig} mm	Δd mm
1500	450	16-35	0.5
	500	40-50	0.5
	500	55-90	0.8
	500	95 or over	1.0
3000	500-650	40-50	0.5
		55-70	0.8
	630-880	75-180	0.8
	750	190 or over	1.2

Figure 5 contains oscillograms recorded during the piercing of billets in containers with dies and (a) cylindrical and (b) tapered mandrel noses. When cylindrical mandrel noses are used the piercing force reaches a specific value and then remains constant until the moment at which the "plug" shears. The result of using tapered mandrel noses, even if the taper is small, is that the curve for the forces on the mandrel nose is of an extreme nature (fig.5b), while the maximum piercing forces are 16-22% greater for copper and up to 50% greater for L62 brass. This must be taken into account, particularly in extruding small diameter tubes, when the strength of the mandrel nose limits the piercing process. In certain cases, obviously it is advisable to use cylindrical mandrel noses.

Effects of the pressure head stress. When a billet has been extruded in a container, in order to save time and muscular effort by the operator, in the case of mandrel noses with diameters less than 100mm the pressure is sometimes not taken off the main plunger and the billet is pierced with a pressure head. The pressure head is the stress p_{podp} exerted on the press plate during piercing, i.e. the force p_{gp} of the main ram related to the transverse cross-sectional area of the pierced billet:

$$p_{podp} = \frac{P_{gp}}{F_k - F_{ig}} \; kg/mm^2 \; (10 \; MN/m^2) \qquad (2)$$

Figure 6 shows the change in the stress in mandrel noses in the initial moment of piercing with relation to the pressure head stress. Altering the pressure head stress from 0 to 150 MN/m^2 (from 0 to 15kg/mm^2) increased the initial stress in the mandrel nose by 40-150%.

The increase in the stress σ_{podp} in the mandrel nose due to the pressure head can be found from the power equation:

$$\sigma_{podp} F_{ig} v_{ig} = p_{podp} (F_k - F_{ig}) \, v_{gp} \qquad (3)$$

whence:

$$\sigma_{podp} = \frac{v_{gp}}{v_{ig}} \cdot \frac{F_k - F_{ig}}{F_{ig}} \; p_{podp} \qquad (4)$$

The ratio of pressure pad (main plunger) speed v_{gp} to mandrel nose speed v_{ig} can be found from the volumes displaced in a time τ:

$$F_{ig} \frac{\ell_{ig}}{\tau} = (F_k - F_{ig}) \frac{\ell_{p.sh}}{\tau} \qquad (5)$$

here ℓ_{ig} and $\ell_{p.sh}$ are the respective travels of the mandrel nose and the pressure pad.

If the packings in the hydraulic system are not well enough fitted, this may cause the pressure pad to withdraw during piercing. Research has proved that the pressure head stress

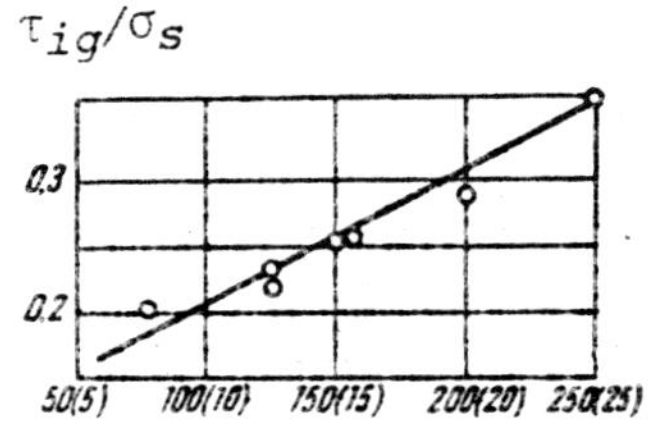

Pressure head stress p_{podp}, MN/m^2 (kg/mm^2)

Fig.7 Effects of the pressure head stress p_{podp} on the frictional stress $\frac{\tau_{ig}}{\sigma_s}$. Horizontal 1500 tonne (150 000kN) hydraulic press. MNZh5-1 alloy, D_k = 155mm, d_{ig} = 35mm.

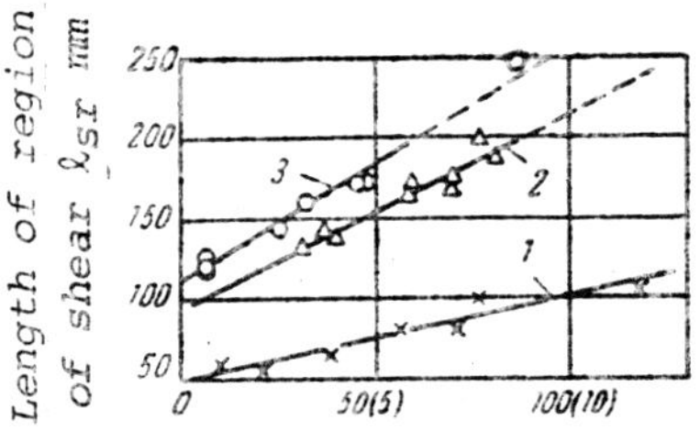

Pressure head stress p_{podp}, MN/m^2 (kg/mm^2)

Fig.8 Relationship of the length ℓ_{sr} of the region of shear to the pressure head stress p_{podp}. 1500 tonne (150 000kN) press. L62 brass, T_{sl} = 720-740°C, D_k = 180mm, mandrel nose dia. d_{ig} : 1-26mm; 2-40mm; 3-50mm.

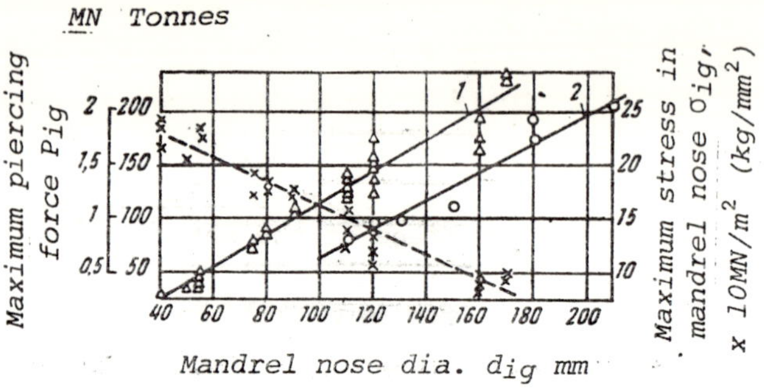

Fig.9 Relationship of the maximum force P_{ig} and the stress σ_{ig} in the mandrel nose during the piercing of a billet to the mandrel nose dia. d_{ig}. 3000 tonne press: (1) copper, billet temperature T_{sl} = 880-920°C, container dia. D_k = 306mm, L_{sl} = 400-450mm; (2) Br.AZhN10-4-4 bronze, T_{sl} = 890-920°C, D_k = 408mm, L_{sl} = 550mm. —P_{ig} ----σ_{ig}

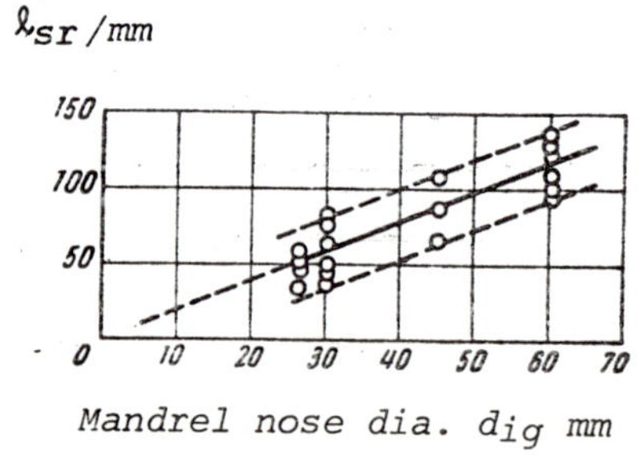

Fig.10 Relationship of the distance between the maximum piercing force and the die (the length of the shear region ℓ_{sr}) to the diameter of the mandrel nose. 1500 tonne press, copper billet temperature 770-800°C. Container dia. D_k = 180mm. Billet length L_{sl} = 240-270mm.

remains practically constant throughout the piercing of a billet. Since:

$$v_{ig} = \frac{\ell_{ig}}{\tau} \text{ and } v_{gp} = \frac{\ell_{p.sh}}{\tau} \text{ then } \frac{v_{gp}}{v_{ig}} = \frac{F_{ig}}{F_k - F_{ig}} \text{ and } \sigma_{podp} = p_{podp} \quad (6)$$

In accordance with equation (6), the increase in the stress in the mandrel nose is proportional to the pressure head stress. The experimental data plotted in fig.6 confirm this. Values of the angular coefficients found by experiments are practically equal to unity.

The effects of the pressure head stress on the tangential frictional stresses over the surface between the mandrel nose and the metal are plotted in fig.7 and assessed in terms of the ratio $\frac{\tau_{ig}}{\sigma_s}$*. Increasing the pressure head stress causes the ratio $\frac{\tau_{ig}}{\sigma_s}$ to increase by linear law. The fact that $\frac{\tau_{tr}}{\sigma_s}$ is proportional to p_{podp} confirms that in this case there is Coulomb friction. The value of the ratio $\frac{\tau_{ig}}{\sigma_s}$ for the MNZh5-1 alloy (see fig.7), is between 0.1 and 0.35. The same limits are also characteristic of copper. If allowance is made for the local chilling of the contact layer of metal by the mandrel nose (increase in σ_s) the true value of the ratio $\frac{\tau_{ig}}{\sigma_s}$ will be even lower.

The effect of the pressure head stress on displacement of the piercing force maximum p_{max} was determined by experiments along the region of shear ℓ_{sr} found from the travel indicator.

We can see from the curves in fig.8 that, as the pressure head stresses increase, so also does ℓ_{sr}. The reason for this is that, as the pressure head increases, the resistance of the metal to flow towards the pressure pad increases, while the magnitude of the forces on the mandrel nose during metal flow in the reverse direction becomes equal to the force of shear earlier.

Effects of mandrel diameter. The principal factor affecting the piercing force is the diameter of the mandrel. Within the range investigated (fig.9), the maximum piercing force increases in direct proportion to the increase in the mandrel diameter. When the diameter of the mandrel was increased from 40 to 170mm, the

---

* Here and henceforth σ_s is the yield stress (strength) of the metal being deformed at the extrusion temperature. The extrusion temperature is taken as being the temperature of the billet when the latter has been heated.

maximum piercing force for copper billets (curve 1 in fig.9) increased from 30 to 220 tons.

The increase in the piercing forces is associated both with increase in the "frontal resistance" (due to increase in the transverse cross-sectional area of the mandrel nose) and to the forces of friction (increase in the area of the contact surface between the mandrel nose and the metal).

As the diameter of the mandrel nose is increased, the frontal resistance also increases, and as a result the maximum piercing force increases and the force necessary for shear deformation is reached earlier. The maximum piercing force shifts towards the press plate (16). This is confirmed by the experimental curve for the relationship of the length ℓ_{sr} of the region of shear to mandrel nose diameter plotted in fig.10.

Effects of container diameter. It was established, on the basis of the research results, that alteration in the forces of friction along the surface of contact between the container and the metal has no appreciable effect on the process of piercing.

Using the power equation we can find the increase in the stress σ_{is} in the mandrel nose caused by the resistance of the forces of friction T_Σ to metal flow:

$$T_\Sigma v_m = \sigma_{is} F_{ig} v_{ig} \tag{7}$$

The forces of frictions T_Σ are equal to:

$$T_\Sigma = \pi \ell_x (d_{ig} \tau_{ig} + D_k \tau_k)$$

here τ_{ig} and τ_k are respectively the forces of friction for the mandrel nose and the container, and ℓ_x is the length of billet pierced - mm.

Since the relationship between the rate of flow v_m in the metal and the mandrel nose speed v_{ig} is governed by the equation:

$$\frac{v_m}{v_{ig}} = \frac{d_{ig}^2}{D_k^2 - d_{ig}^2}$$

if we substitute these values in equation (7), we get:

$$\sigma_{is} = \frac{4\ell_x (d_{ig} \tau_{ig} + D_k \tau_k)}{D_k^2 - d_{ig}^2} \tag{8}$$

Assuming that the specific forces of friction along the container and around the mandrel nose are the same, i.e. that $\tau_k = \tau_{ig} = \tau_{tr}$, we get:

$$\sigma_{is} = \frac{4\ell_x \tau_{tr}}{D_k - d_{ig}} = \frac{4\ell_x a\sigma_s}{D_k - d_{ig}} \qquad (9)$$

Figure 11 shows the experimentally found relationship of the frictional stress $\left(\frac{\tau_{ig}}{\sigma_s}\right)$ to the mandrel nose diameter d_{ig} for the piercing of copper and brass billets in 15 000 and 30 000kN (1500 and 3000 ton) presses. We can see from these graphs that the frictional stress drops as the diameter of the mandrel nose is increased. For instance, when copper billets are pierced, increasing the diameter of the mandrel nose from 26 to 120mm. lowers the ratio $\frac{\tau_{ig}}{\sigma_s}$ from 0.25 and 0.05, i.e. by five times.

Since the metal cools more as the diameter of the mandrel nose is increased, the frictional force is bound to drop; this is confirmed by the relationship which has been given.

The magnitude of the "frontal" resistance σ_0, associated with the work of metal shape change, depends on the parameters of the region of deformation and on the ratio $\frac{d_{ig}}{D_k}$. Most authors associate the effect of the ratio $\frac{d_{ig}}{D_k}$ on σ_0 (1,22,24) with the action of friction along the surfaces of contact between the mandrel nose and the metal, and between the container and the metal, when the latter is flowing in a direction opposite to that of the mandrel nose (enclosed piercing). The relationship $\sigma_0 = \phi\left(\frac{d_{ig}}{D_k}\right)$ obtained by L V Prozorov (23,25) and other authors is of an extreme nature with a minimum lying between $\frac{d_{ig}}{D_k} = 0.65\text{-}0.72$.

Figure 12 shows the relationship of σ_0 to the diameter of the mandrel nose for different container diameters. In this case the ratio $\frac{d_{ig}}{D_k}$ varied between 0.15 and 0.6.

When, therefore, $\frac{d_{ig}}{D_k}$ varies within the limits indicated, the diameter of the container has no appreciable effect on the "frontal" resistance during piercing.

Effects of billet length. During enclosed piercing with cylindrical mandrel noses, the length of the billet has practically no effect on the piercing force. S I Gubkin (1) and I Ya Tarnovskii (24) stated that this is because the parameters of the region of deformation become more stabilized at a particular moment and the length of the zone of contact between

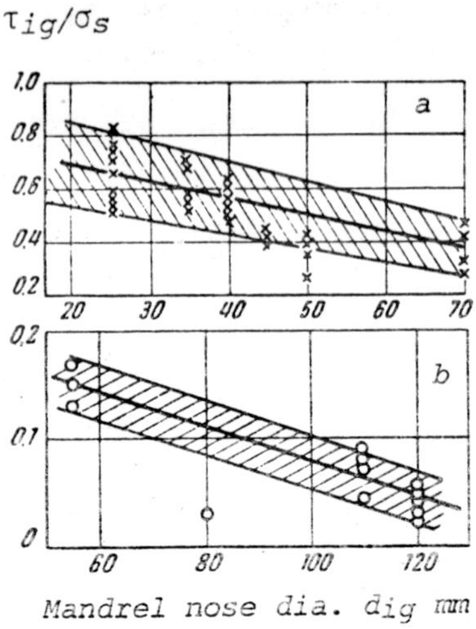

Fig.11

Relationship of the frictional stress $\frac{\tau_{ig}}{\sigma_s}$ to the diameter of the mandrel nose: (a) 1500 tonne press, L62 brass, T_{sl} = 720-750°C, D_k = 180mm; (b) 3000 tonne press, copper, T_{sl} = 840-900°C, D_k = 306mm.

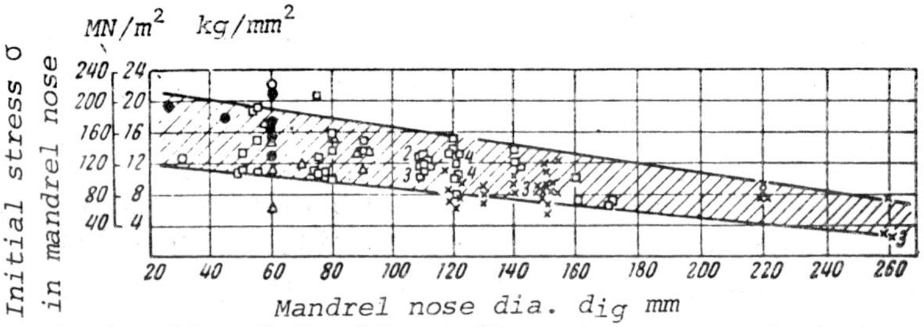

Fig.12 Relationship of the "frontal" resistance σ_0 during piercing to the diameter of the mandrel nose during the extrusion of copper tubes. Pressure head stress p_{podp} = 0. Billet temperature T_{sl} = 850-930°C: (●) D_k = 155mm; (○) D_k = 180mm; (□) D_k = 306mm; (X) D_k = 408mm.

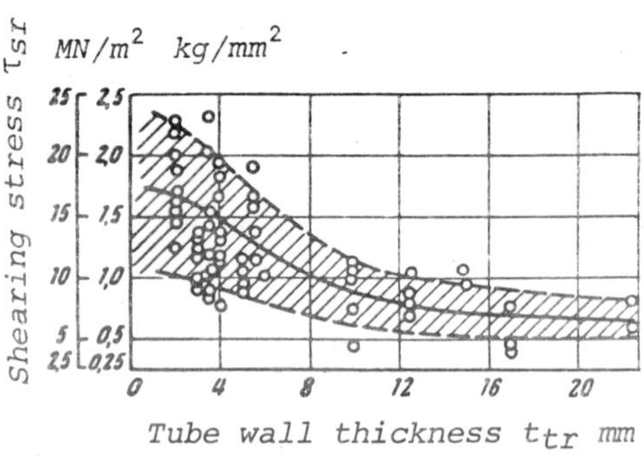

Fig.13

Relationship of the shearing stress τ_{sr} to the tube wall thickness t_{tr}. 1500 tonne and 3000 tonne presses, L62 brass, billet temperature T_{sl} = 720-760°C, container dia. D_k = 155,180 and 306mm.

metal and mandrel nose in which plastic deformation takes place does not alter. This has been confirmed by experiments performed by F I Gofman (29) and L V Prozorov (25).

The situation is different when billets are pierced in containers with tapered mandrel noses. Even an extremely slight nose taper creates a certain definite zone of friction between the mandrel nose and the surface of the metal throughout the entire piercing cycle. The longer the billet, therefore, the greater the force to which the mandrel nose is subjected.

When copper billets are pierced in a 3000 ton press, altering the length of the billet by about 25% increases the piercing force by 15%. As the piercing depth increases, the contact friction forces also increase, and this makes it more difficult for metal to flow in a direction opposite to that in which the mandrel nose moves. As a result, as the length of the billet increases the shear deformation begins earlier. The relative position of the force maximum shifts towards the pressure pad.

Effects of die aperture diameter. In the second stage of piercing the area of shear (see equation (1)) is proportional to the diameter D_{tr} of the die aperture. When, therefore, the diameter of the tube being extruded (its wall thickness) is increased, with d_{ig} = const. the piercing force maximum must be shifted towards the die. As the tube wall thickness is increased, however, the shearing stress τ_{sr} decreases. This is due to the fact that the shear deformations spread in a large volume of metal and with the same absolute deformation ($\Delta \ell_{sr}$ = const.) the thinner the wall of the deformed volume the greater the relative deformation. Thus, as the wall thickness increases the work-hardening of the metal along the shear planes decreases, and in consequence the shearing stresses τ_{sr} decrease (fig.13). As the tube wall thickness is increased from 2 to 10mm the length of the region of shear increases from 150% (fig.14). The reason for this is that, in this case, the reduction in the shearing stresses is the dominant factor.

The experiments which were carried out showed that, as the diameter of the die aperture (i.e. also the tube wall thickness) decreases the shearing stress τ_{sr} changes and the piercing force maximum is shifted. As regards the magnitude of the shearing force, this is practically independent of the die aperture diameter.

Effects of the piercing rate. Research was carried out to establish the effect of the piercing rate on the force. When 34mm diameter copper tubes 2mm thick and 40mm diameter MNZh5-1 alloy tubes 2.5mm thick were extruded in a 1500 ton press (container dia. D_K = 155mm, billet temperatures; copper

T_{sl} = 840-880°C; MNZh5-1 - T_{sl} = 890-920°C), the piercing rate varied between 0.13 and 0.3m/s. The more than 2:1 variation in the piercing rate did not have any appreciable effect on the piercing force which remained practically unchanged. The ratio of τ_{sr} to the yield stress σ_s of the metal found, from the experimental data, for the particular billet temperature concerned also remained unchanged.

Diagram of variation in the piercing force

Figure 15 shows how the stress in the mandrel nose alters when billets are pierced in a container with a die. There are the following characteristic regions on the diagram.

In region I the stress σ_{ig} in the mandrel nose increases from zero to a value σ_0 denoting the start of a steady piercing process. When there is a pressure head stress at the pressure pad side, the stress in the mandrel nose further increases, at the end of region I, by σ_{popd}.

In region II, as the billet is pierced metal flows in a direction opposite to that in which the mandrel nose moves. When the region of deformation has become stable (in the case of cylindrical mandrel noses), the stresses in the mandrel nose do not alter. When tapered noses are used the stress in them increases by an additional amount σ_{tr} by comparison with cylindrical mandrel noses. The stress in the mandrel nose in region I is denoted by the sum:

$$\sigma_{ig}^{II} = \sigma_o + \sigma_{podp} + \sigma_{tr} + \sigma_{is} \qquad (10)$$

The magnitude of the last two summands σ_{tr} and σ_{is} is a function of the length ℓ_x of the pierced part of the billet.

Shear deformations are characteristics of region III; the volume of metal beneath the end of the mandrel nose is shifted (sheared) relative to the remainder of the billet. The stresses in the mandrel nose, therefore, amount to the forces of friction along the nose in the pierced part of the billet and the shearing forces:

$$\sigma_{ig}^{III} = \sigma_{tr} + \sigma_{sr} \qquad (11)$$

Here the "frontal" resistance σ_{sr} is a function of the distance from the end of the mandrel nose to the die and the shearing stress τ_{sr}. In region III the stresses σ_{ig} as a rule decrease.

As the "scalp" forms, the forces of friction preventing the shifted volume of metal from moving into the die increase (region IIIa) until the die aperture has been completely filled; they then remain unchanged (region IIIb).

The total stresses in the mandrel nose in region IIIb are, therefore:

$$\sigma_{ig}^{III} = \sigma_{tr} + \sigma_{sr} + \sigma_{m} \tag{12}$$

where σ_m is the increase in the stresses in the mandrel nose due to the forces of friction in the working band of the die aperture.

In region IV the scalp is separated; in most cases this is accompanied by the formation of a neck (the tube wall ruptures). When thick walled tubes are extruded the forces required to rupture the tube may exceed the piercing forces at the end of region III, and the "scalp" is therefore pierced without being separated.

Analytical determination of the piercing force

Calculation of the maximum piercing force amounts firstly to determining the moment at which shearing begins, and secondly to determining the stress in the mandrel nose corresponding to the end of region II (see fig.15).

Equality between expressions (10) and (11) corresponds to the position of the maximum stresses in the mandrel nose:

$$\sigma_{ig}^{II} = \sigma_{ig}^{III}$$

or

$$\sigma_{0} + \sigma_{podp} + \sigma_{tr} + \sigma_{is} = \sigma_{tr} + \sigma_{sr} \tag{13}$$

Since for this case:

$$\sigma_{is} \approx \frac{4\tau_{tr}(L_{sl} - \ell_{sr})}{D_k - d_{ig}} \tag{14}$$

$$\sigma_{sr} = \frac{P_{sr}}{F_{ig}} = \frac{2(D_{tr} + d_{ig})\,\ell_{sr}\tau_{sr}}{d_{ig}^2} \tag{15}$$

and both the left and right hand sides of equations (13) include σ_{tr}, by solving equation (15) for ℓ_{sr} we find the length of the region of shear:

$$\ell_{sr} = \frac{\sigma_{0} + \sigma_{podp} + \dfrac{4\tau_{tr}L_{sl}}{D_k - d_{ig}}}{\dfrac{4\tau_{tr}}{D_k - d_{ig}} + \dfrac{2(D_{tr} + d_{ig})}{d_{ig}^2}\tau_{sr}} \tag{16}$$

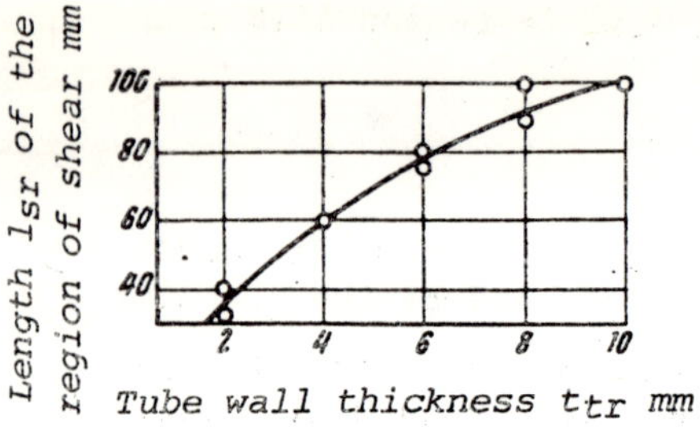

Fig.14
Relationship of the length ℓ_{sr} of the region of shear to the tube wall thickness t_{tr}. 1500 tonne press, D_k= 155mm, copper T_{sl}=850-900°C, mandrel nose dia.d_{ig}=30mm.

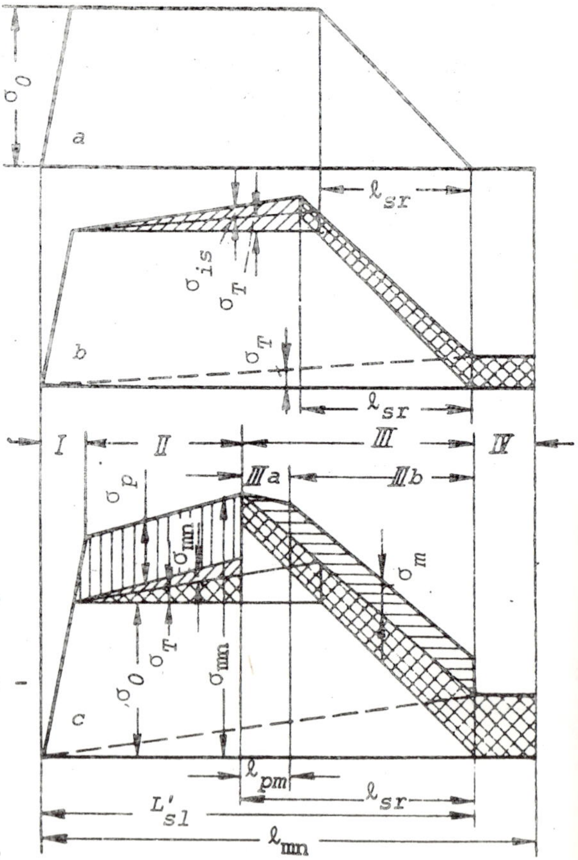

Fig.15
Diagram of variation in the stress in a mandrel nose during the piercing of a billet: (a) with no pressure head stress or forces of friction; (b) with no friction at the mandrel nose - metal and container -metal surfaces; (c) with friction along the mandrel nose, the container and the die aperture surfaces, also with a pressure head (stress in the pressure pad).

When tubes are extruded by the direct method, the ratio $\frac{d_{ig}}{D_k}$ does not exceed 0.5–0.6. For values of $\frac{d_{ig}}{D_k} < 0.5\text{–}0.6$, the value of σ_{is} is only 3–5% of $\sigma_0 + \sigma_{podp}$, is only 3–5%. We can therefore ignore the effect of σ_{is}

Equation (16) can then be greatly simplified:

$$\ell_{sr} = \frac{\sigma_0 + \sigma_{podp}}{\tau_{sr}} \cdot \frac{0.5d_{ig}}{\frac{D_{tr}}{d_{ig}} + 1} \qquad (17)$$

For the particular case of extrusion we can take the ratios $\frac{\sigma_0 + \sigma_{podp}}{\tau_{sr}}$ and $\frac{D_{tr}}{d_{ig}}$ as being constant. The extent of ℓ_{sr} of the region of shear then depends linearwise on d_{ig}; this is confirmed by experimental data.

The values of ℓ_{sr} calculated using equation (17) and found from the oscillograms are given in Table 3.

The maximum stress $(\sigma_{ig})_{max}$ in the mandrel nose during piercing is governed by equations (10) or (11) with $\ell_x = L_{sl} - \ell_{sr}$ and values of σ_{tr} equal to:

$$\sigma_{tr} = 4\tau_{tr} \frac{L_{sl} - \ell_{sr}}{d_{ig}} \qquad (18)$$

It was proved earlier that, for the range considered, the ratio $\frac{\sigma_0}{\sigma_s}$ depends linearwise on the mandrel nose diameter d_{ig}:

$$\frac{\sigma_0}{\sigma_s} = A - Bd_{ig} \approx 10 - 0.025d_{ig} \qquad (19)$$

The values of the constants A and B are found by the statistical analysis of experimental data.

Thus,

$$(\sigma_{ig})_{max} = \sigma_s \left[A - Bd_{ig} + \frac{4_a(L_{sl} - \ell_{sr})D_k}{(D_k - d_{ig})\, d_{ig}}\right] + p_{podp} \qquad (20)$$

or:

$$(\sigma_{ig})_{max} = \frac{2\sigma_s}{d_{ig}} \left\{ 2a(L_{sl} - \ell_{sr}) + \frac{(D_{tr} + d_{ig})\ell_{sr}}{d_{ig}} b \right\} \qquad (21)$$

where: $a = \frac{\tau_{tr}}{\sigma_s}$ and $b = \frac{\tau_{sr}}{\sigma_s}$.

The values of the coefficients a and b are given in Table 4.

When the true and calculated piercing forces are compared (Table 5) we can see that equations (20) and (21) provide results which agree satisfactorily.

Calculation of the endurance of mandrel noses during piercing. When mandrel noses are calculated for longitudinal bending, the following conditions for fixing the ends of the bar are used; one end is rigidly fixed in the mandrel holder, the other pivoted. The critical stress σ_{kr} is in inverse proportion to the square of the free (not entering the billet) length of the mandrel (34):

$$\sigma_{kr} = \frac{\pi^2 E i^2}{\mu^2 \ell^2} \qquad (22)$$

here E is the modulus of elasticity of the metal of which the mandrel nose is made; μ is the coefficient of endurance, equal to 0.7; ℓ is the free length of bar; i is the radius of inertia.

It is advisable to verify the endurances of mandrel noses at the beginning of the process of piercing, with the maximum free length.

On the basis of the margin of endurance n_u the maximum billet length with which the mandrel nose is provided with sufficient resistance to longitudinal bending can be calculated for the start of the piercing process using the equation:

$$L_{sl} = \frac{\pi i}{\mu} \sqrt{\frac{E}{\sigma_s (A - Bd_{ig}) n_u + p_{podp}}} - H_{p.sh} \qquad (23)$$

where $H_{p.sh}$ is the thickness of the press plate mm; n_u is the coefficient of the margin of endurance, $n_u \geqslant 1.0$;

alternatively:

$$L_{sl} \leqslant 1.12 d_{ig} \sqrt{\frac{E}{\sigma_s (10 - 0.025 d_{ig}) n_u + p_{podp}}} - H_{p.sh} \qquad (23a)$$

Table 3

Length of the region of shear, ℓ_{sr}, for piercing billets of MNZh5-1 alloy

Container diameter D_k mm	Tube dimensions $d_{ig} \times D_{tr}$ mm	Length L_{sl} of extruded ingot (from oscillo-grams) mm	Billet Temperature T_{sl} °C	Pressure head stress p_{podp} MN/m^2 (kg/mm^2)	$\frac{\tau_{sr}}{\sigma_s}$	ℓ_{sr} mm	
						From oscillo-grams	By equation (17)
155	60 × 78.5	365-370	900-930	170-190 (17-19)	0.75	330-350	270-297
	30 × 61.5	175-180	900-920	168-228 (16.8-22.8)	0.70	140-170	112-132
	40 × 66.5	175-185	890-930	157-173 (15.7-17.3)	0.70	110-150	159-173
	40 × 48	175-185	890-940	125-236 (12.5-23.6)	0.95	90-170	111-172
	40 × 48	360-375	890-970	157-236 (15.7-23.6)	0.95	120-360	170-212
	40 × 48	270	890-930	157-267 (15.7-26.7)	0.95	120-220	155-202
	40 × 56	270-280	920-930	157-188 (15.7-18.8)	0.75	160-170	177-192
	40 × 51	270-275	910-930	157-220 (15.7-22.0)	0.85	140-150	167-185
306	60 × 122 30 × 122	380-385 380	900-940 940-950	0 0	0.65 0.50	120-125 90-100	129-130 55

Table 4

Values of the coefficients a and b for calculating piercing forces

Metals	a									
	Mandrel nose diameter d_{ig} mm									
	20	30	40	60	80	100	140	160	180	200
Copper	0.25	(0.22)	0.20	0.15	0.10	(0.7)	0.04	0.03	0.02	0.02
L62 brass	0.70	(0.60)	0.56	0.44	0.3	0.18	-	-	-	-
LS59 brass	0.90	(0.70)	(0.50)	(0.3)	(0.2)	0.13	0.07	-	-	-
MNZh5-1	(0.30)	0.24	(0.20)	(0.16)	(0.1)	0.08	(0.06)	0.04	-	-
BrAZhN10-4-4	-	-	-	0.15	0.09	0.04	(0.04)	(0.3)	(0.03)	0.02

Metals	b										
	Tube wall thickness t_{tr} mm										
	2.0	2.5	3.0	4.0	5.0	7.0	10.0	15	20	25	45
Copper	1.2	(1.1)	(1.0)	1.0	(0.9)	(0.8)	0.7	0.6	0.5	0.5	0.5
L62 brass	1.6	(1.5)	(1.4)	1.3	(1.1)	(0.9)	0.78	0.65	0.55	0.5	-
LS59 brass	-	-	(1.8)	1.6	(1.4)	(1.2)	1.0	0.71	0.70	-	-
MNZh5-1	-	3.0	1.5	0.95	0.87	0.75	(0.7)	(0.7)	(0.7)	0.65	0.5
BrAZhN10-4-4	-	-	-	-	-	-	-	0.5	-	0.4	-

Note: The figures in parentheses are approximate values

THE FORCES ACTING ON THE MANDREL NOSE DURING EXTRUSION

Research into the forces exerted on the mandrel nose while the metal is flowing into the die is interesting from the point of view of calculating mandrel nose strength and studying the effect of the forces of friction in the region of contact between the mandrel and the metal on the total extrusion force.

Variation in the force exerted on the mandrel nose

The forces acting on the mandrel nose during the extrusion of tubes are governed by the forces of friction between the metal being extruded and the side surface of the mandrel.

In reference (16) the authors of this book proved that the sign of the stresses in the mandrel and the length of the region of the mandrel along which the metal slides during the extrusion of tubes depend on the ratio of the velocities of the mandrel and the press plate. There are the following possible variations of the velocity ratio:

1) the mandrel is stationary relative to the container. An example of this is extrusion with a staged taper ("bottle-shaped") mandrel (5). With this method of extrusion the length of the region in which the metal slides over the mandrel is maximum and equal to the length of the billet;

2) the mandrel is fixed in a holder and cannot move relative to it. In this case the length of the region in which the metal slips over the surface depends on the nature of metal flow;

3) a special design mandrel is so fixed that it can move relative to the holder.

In the case of horizontal hydraulic presses with an independent piercing system, when cylindrical mandrels are used the ratio of the mandrel and holder (press plate) velocities corresponds to the second variation. In most cases, when tubes are extruded, the mandrel plunger cross-member rests against the adjusting nut of the main plunger piston rod, and the velocity of the mandrel is equal to that of the main plunger. This enables us to consider the mandrel as being rigidly fixed to the holder.

The length of the region of slip is, however, affected not only by the method of fixing the mandrel but also by the manner in which the metal flows. Experimental study of the rheology of metal during extrusion (1,3,5,35-37 etc) has proved that the manner of flow can conditionally be subdivided into two extreme cases (4):

a) laminar ("in layers") metal flow, in which the deformation zone is concentrated in the immediate vicinity of the die. The billet moves in the container without being deformed;

Table 5 Initial and maximum piercing forces P_0 and P_{max}

Container diameter D_k mm	Tube dimensions $d_{ig}xD_{tr}$ mm	Length of back-extruded billet L_{sl} mm	Billet Temperature T_{sl}°C	Pressure head MN/m^2
				M1 and M3 coppers
180	30x42	208-242	810-870	70-160
	40x47	294-319	900-950	15-200
	40x50	275	920-950	127-195
	45x51	228-252	932-948	60-90
	60x66	280-412	880-940	0-40
306	110x123	370-410	850-900	0
	120x145	400-510	870-910	0
	110x120	460-480	890-920	0
408	150x230	525-540	890-930	0
	120x170	490-530	870-920	0
	130x220	550-620	900-930	0
	220x240	480-485	820-890	0
				L62 Brass
180	26x33	215-251	770-780	39-193
	26x34	164-225	770-790	20-145
	26x37	186-270	760-780	19-116
	30x60	257-265	760-800	65-108
	35x42	162-222	680-780	50-86
	35x45	213-233	740-780	60-113
	35x60	231-363	770-780	59-98
	40x46	203-231	680-840	0-113
	40x47	307-318	750-820	35-90
	40x60	325-337	730-760	15-95
	45x52	303-340	750-800	26-81
	50x56	299-365	804-824	10-51
	60x66	322-348	800-830	32-117
	70x76	316-335	770-790	17-78
				MNZh5-1 Alloy
155	60x78.5	365-370	900-930	87-190
	30x61.5	175-380	900-910	213-320
	40x66.5	175-185	890-930	157-173
	40x48	175-370	890-950	125-267
	40x56	270-280	920-930	157-180
	40x51	270-275	910-930	157-220
	40x44	270-275	970-980	188
306	60x122	380-385	900-940	0
	30x122	380	940-950	0
	190x204	230-475	850-960	0

stress p_{podp}	Piercing force kN (10^{-1} tons)			
	True		Calculated by equation(20)	
kg/mm²	P_0	P_{max}	P_0	P_{max}
7-16	-	386-460	185-240	248-308
1.5-20	-	364-460	199-420	314-465
12.7-19.5	-	368-420	330-440	376
6-9	306-372	425-442	345-392	410-440
0-4	370-650	500-720	395-651	541-692
0.0	1150	1170-1460	1235-1380	1340-1510
0.0	-	1050-1870	1380-1535	1450-1770
0.0	880-1000	1290-1345	1170-1235	1340-1470
0.0	935-1520	1580-1990	1810-1970	1880-2060
0.0	-	1110-1580	1340-1500	1460-1640
0.0	-	1930-1990	1460-1520	1585-1740
0.0	2340-2630	2570-3330	3080-3590	3190-3690
3.9-19.3	84-202	135-228	62-144	154-202
2.0-14.5	76-161	152-254	50-114	120-182
1.9-11.6	76-118	152-234	80-103	143-190
6.5-10.8	30-53	158-228	98-134	126-165
5-8.6	53-116	188-295	110-222	180-287
6-11.3	60-93	206-239	127-192	212-229
5.9-9.8	175-295	268-364	132-164	144-200
0-11.3	56-159	206-350	170-324	187-380
3.5-9	160-346	253-373	107-148	177-236
1.5-9.5	48-145	228-364	151-228	215-268
2.6-8.1	242-390	298-396	156-256	213-322
1-5.1	130-260	278-335	131-203	190-237
3.2-11.7	410-725	630-800	259-475	323-505
1.7-7.8	260-428	584-890	239-587	309-625
8.7-19	745-1000	1030-1320	820-1132	1045-1250
21.3-32	243-258	314-382	281-382	328-522
15.7-17.3	258-430	445-600	282-487	445-525
12.5-26.7	314-630	460-730	421-581	450-692
15.7-18.8	388-450	540-600	455-495	528-554
15.7-22	430-450	580-625	444-545	522-585
18.8	450	560-580	437-460	440-488
0.0	-	655-720	509-602	780-840
0.0	-	328-394	130-137	292-306
0.0	2100-3430	3590-4400	2840-4900	-

b) turbulent ("zonal") flow, in which owing to the contact friction conditions and the temperature gradient throughout the billet deformation of the metal takes place throughout its entire volume except for the "dead" zones against the pressure pad and the die.

In view of this, when tubes are extruded the length of the billet can be subdivided into five regions in which the metal exerts different actions on the mandrel nose.

Region I - the "dead" zone against the pressure pad, in which the velocity of the metal relative to the mandrel is close to zero.

Region II - in which during the laminar flow the velocity with which the metal slips over the mandrel is relatively constant and close to zero, while with turbulent flow it is not zero and its vector is towards the die.

Region III - this is the geometrical region of deformation immediately adjacent to the die. In this region the velocity of the metal relative to the mandrel is maximum, and varies between zero and $v_{g.p}$ $(\mu-1)$; here μ is the total reduction and $v_{g.p}$ is the velocity of the main plunger (the mandrel).

Region IV - the length of which corresponds to that of the forming band in the die. The velocity of the metal relative to the mandrel is constant and equal to $v_{g.p}$ $(\mu-1)$

Region V - governed by the length of the mandrel nose projecting beyond the forming band of the die. The velocity of the metal relative to the mandrel is the same as in region IV.

With laminar metal flow a tensile force is exerted on the mandrel nose in regions III-V; this is partially counteracted by the forces of friction of rest in regions 1-2. With turbulent flow a tensile force develops in regions II-V, and is partially counteracted in region I. Thus the force absorbed by the mandrel holder is the difference between the forces of friction developed in regions in which the velocity of the metal relative to the mandrel differs from zero and the forces of friction of rest in the regions in which the velocity of sliding is equal to zero.

The maximum tensile force on the mandrel in turbulent metal flow therefore corresponds to the metal beginning to flow into the die, when the total length of region II and III is greatest.

Analysis of oscillograms proved that the extrusion of tubes made of L62 brass, MNZh5-1 alloy and MNZhMts30-0.8-1 cupronickel corresponds to this case; the region of deformation covers practically the entire length of the billet, and amounts to 80-90% of the length of the back-extruded billet (Table 6).

Table 6

Length of the region of deformation during the extrusion of tubes from a container with a diameter D_k = 180mm

Metal	Tube dimensions $d_{ig} \times D_{tr}$ mm	Length of back-extruded billet L_{sl} mm	Length $\ell_{o.d}$ of the region of deformation mm	
			From oscillograms	By equation (24)
Copper	26 x 33 30 x 37 40 x 47	250-260 250-260 230-240	110-115 95-116 100-105	91.9 91.0 88.6
L62 brass	26 x 33 26 x 34 30 x 37	245-260 210-220 250-260	210-240 190-205 230-245	- - -
MNZh5-1 alloy	35 x 41*	240-250	210-225	-
MNZhMts30-0.8-1	36 x 52*	230-140	200-210	-

* Container diameter D_k = 155mm

In laminar flow, the length of the region of deformation (region III) is relatively constant, and the longer this zone the more effectively are the tensile forces developed in this zone and in region IV compensated by region II. The maximum of the tensile forces in the mandrel holder therefore corresponds to the moment at which the running length is equal to the length of the region of deformation. As the pressure pad continues to move the tensile stress decreases; this is due to reduction in the length of the region of deformation.

With laminar flow the true length of the region of deformation found from the oscillograms (see Table 6) corresponds well to the length calculated using I L Perlin's equation (4):

$$\ell_{o.d} = \frac{D_k}{2 \sin \alpha} - \frac{D_{tr}}{2 \tan \alpha} \qquad (24)$$

here α is the angle of taper of the die from the axis of the mandrel.

Our research has proved, however, that the length of the region of deformation is greatly affected by the thickness of the tube wall (fig.16). This is obviously due to reduction in the hydrostatic pressure at the die aperture side.

Table 7

Maximum tensile forces acting on the mandrel during the extrusion of copper tubes from 150 x 240mm dia. billets with laminar and turbulent metal flow

Tube dimensions $d_{ig} \times D_{tr}$ mm	Billet temperature T_{s1}	Force P_{ig} on mandrel *MN* (tons)	
		Laminar flow	Turbulent flow*
30 x 38	800-840	0.068-0.072 (6.8-7.2)	0.11-0.126 (11.0-12.6)
30 x 37	820-850	0.065-0.07 (6.5-7.0)	0.105-0.110 (10.5-11.0)

* This is recorded in the direct extrusion of copper immediately after L62 or L68 brass.

It should be noted that, in the case of copper, depending on the contact friction conditions at the surface of the container and the mandrel the form of metal flow may vary from laminar to turbulent, with deformation covering the entire length of the billet. For

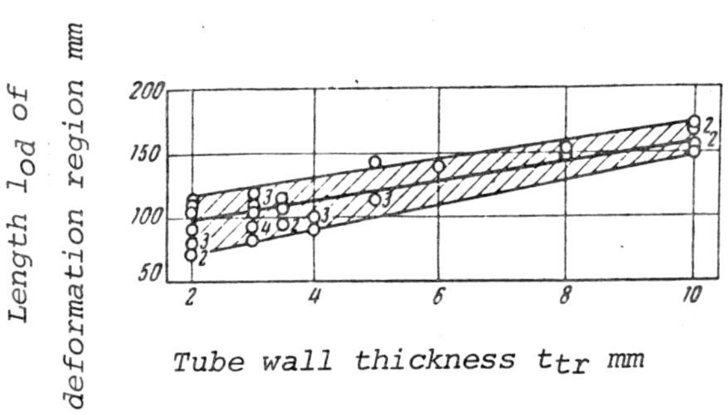

Fig.16 Relationship of the length of the plastic zone (the deformation region) to tube wall thickness for the extrusion of copper tubes. Container dia. D_K = 155mm, mandrel diameter d_{ig} = 30mm, billet temperature t_{sl} = 840-890°C.

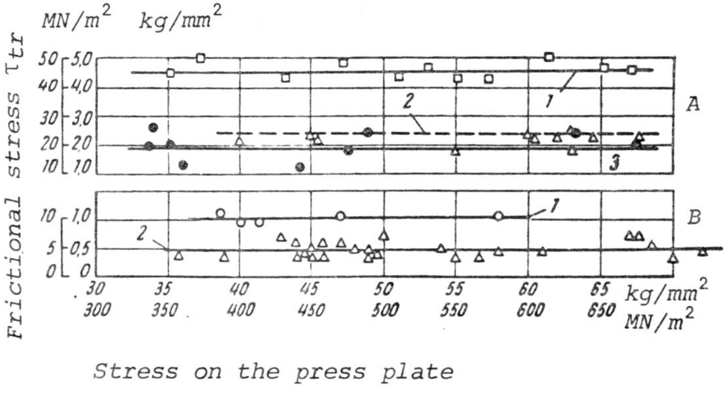

Fig.17 Relationship of the frictional stress on the side surface of the mandrel to the stress on the press plate during the extrusion of: (A) L62 brass D_K = 180mm T_{sl} = 700-750°C. Mandrel diameters d_{ig} : (1) 70mm (2) 35mm (3) 26mm; (B) copper D_K = 155mm T_{sl} = 820-880°C. Mandrel dia. d_{ig}: (1) 60mm (2) 30mm.

instance, the extrusion of copper tubes immediately after brass has been extruded means that there will be "zonal" (turbulent) flow, due to brass having adhered to the container and the mandrel and to steeper temperature gradient from the container into the billet. In this case the force exerted on the mandrel is increased by about 60% (Table 7).

Frictional stress along the side surface of the mandrel

When the experimental data were analysed, the frictional stress τ_{tr} was determined as the ratio of the reduction ΔP_{ig} in the tensile forces acting on the mandrel holder (beyond the maximum on the curve for variation in the forces acting on the mandrel) to the reduction ΔS in the mandrel contact surface area when the length of the billet is reduced by ΔL_{sl}:

$$\tau_{tr} = \frac{\Delta P_{ig}}{\Delta S} = \frac{\Delta P_{ig}}{\pi d_{ig} \Delta L_{sl}} \qquad (25)$$

The effects of the different parameters of the process on the magnitude of τ_{sr} were established on the basis of research into the process.

Effects of stress on the pressure pad. Figure 17 show the relationship of the frictional stress τ_{tr} to the stress on the pressure pad. The specific force of friction is practically independent of the extruding pressure (the stress in the pressure pad) between 250 and 650MN/m^2 (25 and 65kg/mm^2) when L62 alloy or copper tubes are extruded. This confirms I L Perlin's argument, stated in reference (4), that with the high hydrostatic pressures characteristic of the extrusion process it is best to take the forces of friction into account by the law of the maximum tangential stresses, not by the Amonton-Coulomb law. In this case friction is accompanied by shear deformations in the contact layers of the metal, and the magnitude of the specific forces of friction is governed by the mechanical properties of the metal.

In reference (24), I Ya Tarnovskii et alia, noted that, to a great extent the physical state in the contact layers of the metal may differ from that of the principal part of the billet owing to cooling and additional work-hardening. The degree to which the contact layer is cooled depends on the thermal conductivity of the metal being extruded: the higher this conductivity, the less will be the cooling. The ratio $\frac{\tau_{tr}}{\sigma_s}$ therefore alters between 1 and 5 as the stress on the pressure pad is increased when tubes made of L62 brass are extruded (coefficient of thermal conductivity [$\lambda = 196$ W/(m x °C) ($\lambda = 170$kcal/h•m•°C)] it varies between 0.28 and 0.67 when copper tubes are extruded [$\lambda = 340$ W/m•°C ($\lambda = 290$kcal/h•m•°C)].

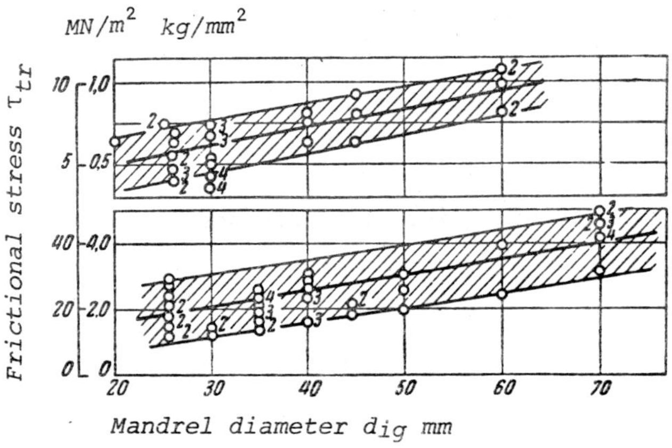

Fig.18 Relationship of the frictional stress τ_{tr} at the side surface of the mandrel to its diameter during the extrusion of different tubes: (a) copper T_{sl} = 880-930°C. Extrusion rate v_{pr} = 0.055-0.090m/s; (b) L62 brass T_{sl} = 880-930°C v_{pr} = 0.035-0.070m/s.

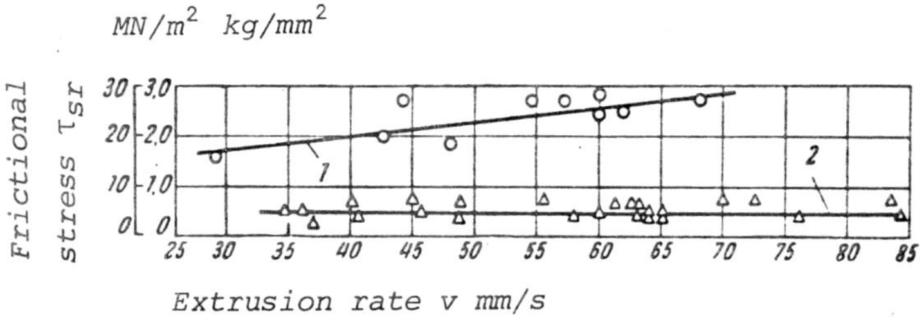

Fig.19 Relationship of the frictional stress on the mandrel to the extrusion rate (the main plunger speed): (1) L62 brass, d_{ig} = 40mm, T_{sl} = 720-760°C, D_k = 180mm; (2) copper d_{ig} = 30mm, T_{sl} = 820-880°C, D_k = 155mm.

Effects of mandrel diameter. The diameter d_{ig} exerts a considerable effect on the cooling of the contact layer. As d_{ig} is increased, the contact layers on metal cool more and the frictional stress increases (fig.18). When L62 brass tubes are extruded, altering the diameter of the mandrel from 26 to 70mm increases the specific forces of friction from 18 to 41MN/m^2 (from 1.8 to 4.1kg/mm^2), i.e. by 130%. When copper tubes are extruded, increasing the diameter of the mandrel from 26 to 60mm increases the specific forces of friction from 4.5 to 10MN/m^2 (from 0.45 to 1kg/mm^2), i.e. by 120%.

Effects of extrusion rate. Figure 19 shows the relationship of the frictional stresses on the surface of the mandrel to the extrusion rate (the speed at which the press plate moves). When grade L62 brass tubes are extruded, as the extrusion rate is increased from 30 to 70mm/s the frictional stress τ_{tr} increases by 50-100%. When copper tubes are extruded, if the extrusion rate is increased from 35 to 85mm/s the frictional stress remains practically constant.

Varying the extrusion rate within the range used in practice for extruding copper tubes thus does not affect the frictional stress. When brass, which has a lower coefficient of thermal conductivity, is extruded the extrusion rate has an indirect effect: with high extrusion rates, the mandrel cannot heat up, the layer of metal in the contact region cools more, and the yield stress of the contact layer of the metal increases. This raises the shearing stress τ_{sr}, and consequently also the tensile force applied to the mandrel.

Effects of billet temperature. As the billet temperature T_{sl} is increased, the frictional forces along the side surface of the mandrel decrease (fig.20). For instance, when copper or L62 brass tubes are extruded, raising the billet temperature by 75-100°C lowers the frictional stress by about 17-29%.

Relationship of the stresses in the mandrel to its diameter

When tubes are extruded, a knowledge of the stresses in the mandrel makes it possible to carry out strength calculations and determine rational technologies. If we ignore the effect of the forces of friction in the mandrel region behind the die aperture (region V), since tapered mandrels are used, also the "dead" zone against the press plate (region I) because it is small, the maximum tensile force and the stress in the mandrel can be determined directly from oscillograms.

Increasing the diameter of the mandrel from 20 to 60mm, i.e. trebling it, with a tube wall thickness t_{tr} = 2-4mm, results in the tensile force increasing, in the case of the extrusion of copper from 60 to 190kN (from 6 to 19 tons), i.e. it is also almost trebled (fig.21). This is due to the increase in the side surface area of the mandrel and in the frictional stress. When the diameter of the mandrel is increased, the tensile force alters linearwise, while the transverse cross-sectional area of the mandrel is proportional to the square of its diameter; the tensile stresses in the mandrel therefore drop as its diameter is increased:

$$(\sigma_{ig})_{max} = \frac{P_{ig}}{F_{ig}} = \tau_{tr} \frac{\ell_{o.d.}}{d_{ig}} \quad (26)$$

Increasing the diameter of the mandrel within the range of diameters given above therefore reduces the tensile stresses by 67% when copper tubes are extruded (see fig.21): from 240 to $80MN/m^2$ (from 24 to $8kg/mm^2$).

The manner in which the tensile force and the stresses in the mandrel alter as the diameter of the latter is changed is similar in the case of extruding L62 brass (see fig.21). In the case of the extrusion of copper tubes, when the tube wall thickness d_{tr} is increased from 2 to 10mm the tensile forces increase from 100 to 150kN (from 10 to 15 tons) (see fig.22). The specific forces of friction on the surface of the mandrel not being related to the stresses on the pressure pad (see fig.17), i.e. to the degree of deformation, the increase in the tensile force is associated with increase in the length of the region of deformation. Actually in this case the length of the region of deformation increased from 100 to 150mm (see fig.16), i.e. by 50%.

Analytical determination of the tensile force on the mandrel during extrusion

The maximum tensile forces acting on the mandrel during the extrusion of tubes can be determined using the following equations:

a) with laminar metal flow:

$$P_{max} = \pi d_{ig} (\ell_{iii}\tau_{iii} + \ell_{iv}\tau_{iv}) \quad (27)$$

b) with turbulent metal flow:

$$P_{max} = \pi d_{ig} (\ell_{ii}\tau_{ii} + \ell_{iii}\tau_{iii} + \ell_{iv}\tau_{iv}) \quad (28)$$

If we ignore the effect of the "dead" zone (region I) and the forces of friction along the mandrel beyond the sizing band of the die (region V) and mean the values of the frictional stress along the mandrel, equations (27) and (28) can accordingly be reduced to the following form:

$$P_{max} = \pi d_{ig} \ell_{sk} \tau_{tr} \quad (29)$$

or:

$$P_{max} = \pi d_{ig} \ell_{sk} a \sigma_s$$

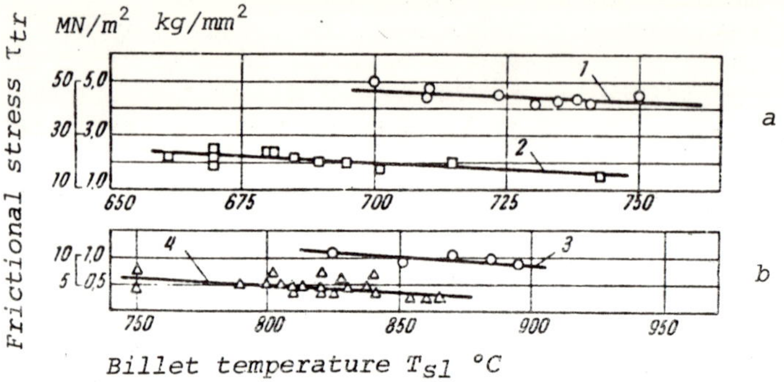

Fig.20 Relationship of the frictional stress along the mandrel, τ_{tr}, to the billet temperature, T_{sl}, for the extrusion of (a) L62 brass and (b) copper tubes: (1) d_{ig} = 70mm, D_k = 180mm; (2) d_{ig} = 35mm, D_k = 155mm; (3) d_{ig} = 60mm, D_k = 180mm; (4) d_{ig} = 30mm, D_k = 155mm.

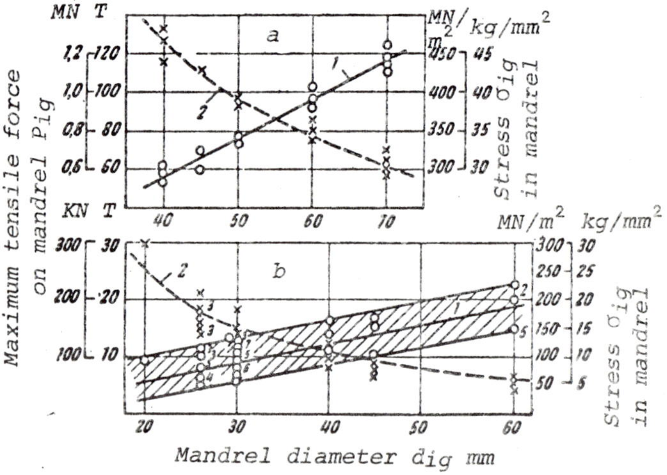

Fig.21

Relationship of (1) the maximum tensile forces and (2) the maximum stresses, in the mandrel, to its diameter: (a) L62 brass T_{sl} = 720-760°C, D_k = 180mm; (b) copper T_{sl} = 820-920°C, t_{tr} = 2-4mm, extrusion rate v_{pr} = 40-70mm/s

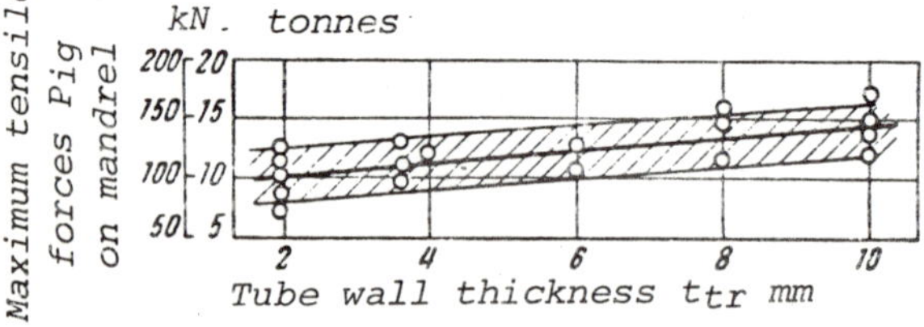

Fig.22 Relationship of the maximum tensile force acting on the mandrel to the tube wall thickness t_{tr} for the extrusion of copper. T_{sl} = 840-890°C (1110-1160°K) D_k = 155mm, mandrel dia. d_{ig} = 30mm.

here ℓ_{sk} is the length of the region in which active forces of friction are effective:

with laminar flow:

$$\ell_{sk} = \ell_{o.d.} + \ell_{p.m.}$$

with turbulent flow:

$$\ell_{sk} \approx 0.8L_{sl} + \ell_{p.m.}$$

here $\ell_{o.d.}$, $\ell_{p.m.}$ and L_{sl} are respectively the length of the region of deformation found using equation (24), the length of the sizing band of the die, and the length of the extruded billet, equal to:

$$L_{sl} = L_{zag} \left(\frac{D_{zag}}{D_k}\right)^2 \qquad (30)$$

In view of the relationship of frictional stress to many factors, according to references (4,24) it is best to assess τ_{tr} as a proportion of the yield stress σ_s of the metal extruded:

$$C = \frac{\tau_{tr}}{\sigma_s}$$

The values of the coefficient C for different metals and alloys are given in Table 8.

Table 8

Values of the coefficient C

Metal	Mandrel diameter d_{ig}, mm						
	20	30	40	50	60	70	180
Copper	0.40	0.45	0.5	0.55	0.6	-	-
L62 brass	1.50	1.70	2.10	2.50	2.90	3.30	-
MNZh5-1	-	0.40	0.50	-	0.6	-	0.8
MnZhMts30 0.8-1	-	0.4	0.5	-	-	-	-

We can see from Table 9, which contains the true and calculated values of the maximum tensile force P_{ig}, that equation (29) provides satisfactory convergence: the deviations do not exceed ±15%.

Calculation of mandrel strength during extrusion

The procedure given in reference (95) is used for obtaining an analytical relationship.

Owing to the radial compression and axial tension, the relative deformation δ_{ig} of the mandrel along its axis should be determined on the basis of the generalized Hooke's law:

Table 9

Comparison of the calculated and true maximum tensile forces on the mandrel during the extrusion of copper tubes

Tube dimensions $d_{ig} \times D_{tr}$ mm	Billet dimensions $D_{zag} \times L_{zag}$ mm	Billet temperature t_{sl} °C	Maximum tensile force on mandrel P_{ig} kN(tons)	
			Calculated	True
26 × 33	175 × 250	800-820	76- 79 (7.6- 7.9)	80-110 (8.0-11.0)
26 × 36	172 × 250	800-820	83- 86 (8.3- 8.6)	105-130 (10.5-13.0)
30 × 37	175 × 250	800-820	104-114 (10.4-11.4)	120-130 (12.0-13.0)
30 × 42	175 × 250	800-820	118-130 (11.8-13.0)	135-150 (13.5-15.0)
40 × 46	175 × 250	850-890	153-166 (15.3-16.6)	140-160 (14.0-16.0)
45 × 51	175 × 250	850-890	172-187 (17.2-18.7)	170-190 (17.0-19.0)
60 × 66	175 × 350	850-900	210-226 (21.0-22.6)	200-230 (20.0-23.0)

$$\delta_{ig} = \frac{1}{E} [\sigma_{ig} + 2\nu\sigma_r]$$

where:

E = the modulus of elasticity of the metal of which the mandrel is made;

ν = Poisson's ratio and is equal to 0.3-0.33;

σ_{ig} = the stress in the mandrel;

σ_r = represents the radial stresses in the mandrel.

Assuming that the elongation deformation is passive (4), we can determine the stresses σ_r from the ductility condition:

$$\sigma_r = P + \sigma_s$$

where P is the longitudinal stress (the mean stress on the pressure pad). (It was assumed in reference (95) that $\sigma_r = P - \sigma_s$).

The mandrel will then begin to rupture (a neck will form in it) under the condition:

$$\sigma_{0.2} \leqslant \sigma_{ig} + 2\nu\sigma_r$$

or:

$$\sigma_{0.2} \frac{4P_{ig}}{\pi d^2{}_{ig}} + 0.6\ (P + \sigma_s) \qquad (31)$$

here: $\sigma_{0.2}$ = the nominal yield stress of the mandrel metal at the extrusion temperature.

If we specify the mandrel strength margin factor k, equation (31) can be used for determining the diameter d_{ig} of the mandrel with which it will be sufficiently durable:

$$d_{ig} \geqslant \frac{4\tau_{tr}\ell_{sk}}{\frac{\sigma_{0.2}}{k} - 0.6\ (P + \sigma_s)} \qquad (32)$$

or:

$$d_{ig} \geqslant \frac{4c\ell_{sk}}{\frac{\sigma_{0.2}}{k\sigma_s} - 0.6\ (1 + \frac{P}{\sigma_s})} \qquad (33)$$

It should be noted that the tensile strength $\sigma_{0.2}$ of the mandrel metal depends to a great extent on the temperature to which the mandrel is heated during extrusion. This in turn is governed by the diameter of the mandrel, also by the duration of the different parts of the extrusion cycle and the method by which the mandrel is cooled.

TEMPERATURE-SPEED CONDITIONS FOR THE EXTRUSION OF TUBES

The degree of deformation, the temperature to which the billet is heated, and the extrusion speed, are the principal technological parameters governing the power of the press and the quality of the product. All these parameters are interconnected, and must therefore be considered in conjunction. The degree of deformation and the temperature of the billet govern the extruding force, and in the case of hydraulic presses with pump and accumulator drives the extrusion speed and the duration of the process depend on this force. The deformation time and the temperature of the billet in turn govern the degree of cooling, and consequently also the temperature of the tube when it leaves the die.

Procedure for measuring the temperature and speed parameters of the process

A type ET-7/110 tachodynamo was used for measuring the speed at which the main plunger and mandrel move. Figure 23 shows the design of the speed sensing element. The pulley 1, which is rigidly connected to the armature of the tachodynamo 2, is rotated by the forward movement of the push rod 3. These push rods are tubes one end of which was fixed to the moving cross-member of the main plunger and the mandrel plunger. To prevent the push rod tube from sliding against the pulley, it is pressed against the latter by means of the springs 4 and the guide rollers 5. The tachodynamo voltage is fed to the oscillograph circuits and recorded on film. Figure 24 shows the general appearance of the speed sensing elements mounted on a press. To obtain the required sensitivity, also to diminish the effects of fluctuations in the resistance of the tachodynamo armature, an additional manganin resistance is connected in series with the armature. The ratio of the resistance of the armature to the additional resistance is 0.06-0.02. The tachodynamo was calibrated before the equipment was used.

In order to determine the travel of the main plunger during extrusion and of the mandrel during piercing, we used the travel indicator shown in fig.23. The contacts of the travel indicator 6 are made of silver with textolite as the insulating material. The textolite half-ring is rotated by the pulley 1 through the reduction gearing 7, which had a transmission ratio of $i = 9.5$. The diameter of the pulley and the transmission ratio were calculated to be such that the length of a pulse on the film corresponded to 20mm of main plunger or mandrel travel.

The temperature of the billet was measured using a fixed position FEP-3 photoelectric pyrometer and an OPPIR-09 optical pyrometer. Since the fixed position FEP-3 pyrometer was located at a certain distance from the container, a high-speed photoelectric pyrometer designed by A V Panov was used for measuring billet temperature

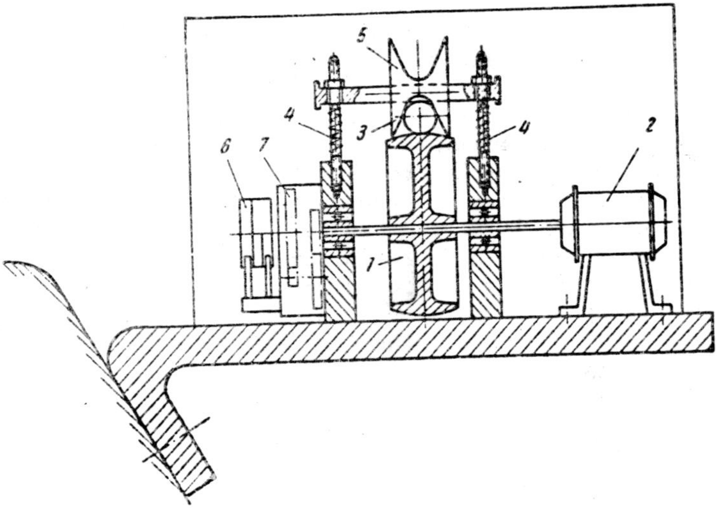

Fig.23 Design of the speed sensing element.

Fig.24 General appearance of the speed sensing element mounted on a press: (1) pulley; (2) tachodynamo; (3) push rod.

immediately before billets entered the container; this made it possible to record the temperature of the billet on the oscillograph film. An equivalent type of pyrometer was used for measuring the temperature of the metal leaving the die. In principle the pyrometer designed by A V Panov is a photopyrometer with the beam of light modulated by means of a mechanical obturator. Figure 25 shows the general appearance of the apparatus. The photopyrometer consists of a head and a control panel with a power supply amplifier. The head includes the optical system, the obturator in the form of a slotted disc which is rotated by a single-phase synchronized motor providing a modulation frequency of 3000c/s, a TsV-3 vacuum element (or FSA1 photoresistor), an amplification cascade adjusted to a frequency of 3000c/s, and an output cathode repeater enabling the head to be taken 15-20m from the control panel. The control panel houses a resonance amplifier with a demodulator and an output d.c. amplifier, to the cathode circuit of which the oscillograph circuit is connected. The pyrometer is so designed that it can record temperature changes taking place at 50 000°C/s.

The photopyrometer irradiation angle is very small, about 15 minutes, and this makes it possible to work from a very small sighting area of the order of 0.75mm^2. The distance of the body being investigated from the photopyrometer object glass is fixed at twice the focal length. The sensitivity of the photopyrometer was varied by means of an electrical potentiometer in the resonance amplifier grid circuit.

Photoelectric pyrometers are partial radiation pyrometers owing to the selectivity of the spectral sensitivity of the photoelectric element or photoresistance. The temperature of a body is therefore measured from the intensity of radiation within a specific range in the radiation spectrum for an incandescent body (42). Radiation from a body will depend on its blackness, and therefore in order to obtain reliable results the method used for calibrating the pyrometer was very important.

In our case calibration was carried out by means of FEP-3 and OPPIR-09 pyrometers, also with a chromel-alumel thermocouple mounted in the press base. The pyrometer for measuring the temperature of the metal leaving the die was so mounted in the press that it could be moved in the vertical plane in order to maintain the required distance between the object glass of the pyrometer and the surface of the metal when the diameter of the extruded tube was changed. The pyrometer for measuring the temperature was mounted on the forward housing of the press, 1445mm from the die. The extent to which the tube metal cooled owing to radiation and convection while it was passing along this distance was calculated by the procedure given in reference (43).

Similar calculations were carried out to determine the temperature drop in the billet between the moment at which it was inserted into the heating device and the moment at which it entered the container. In this case also, with the exception of anomalous deviations from working rythm, the temperature drop in the billet was small.

Speed characteristics of presses with independent piercing systems

In reference (53), Gagen-Torn proved that the relationship of the pressure to the extrusion speed is governed by the equation (the speed characteristics):

$$p_{gp} = p_{akk} - \xi_{gp} v^2_{gp} \qquad (34)$$

here:

p_{gp} and p_{akk} equal the pressures of the liquid in the main ram cylinder and the high pressure accumulator respectively;

v_{gp} = the speed of travel by the main cylinder;

ξ_{gp} = the reduced coefficient of hydraulic resistance of the system consisting of the accumulator apparatus and the main plunger cylinder.

Equation (34) was deduced on the basis of the assumption that the resistance to deformation does not depend on the extrusion speed when the latter is altered over the practical value range.

For presses with independent piercing systems the speed charactertistic takes the similar form:

$$p_{ig} = p_{akk} - \xi_{ig} v^2_{ig} \qquad (35)$$

here:

p_{ig} = the pressure of the liquid in the mandrel cylinder in bars (atm);

v_{ig} = the mandrel plunger speed in m/min;

ξ_{ig} = the reduced coefficient of hydraulic resistance for the system consisting of the accumulator plant and the mandrel cylinder.

Speed characteristics for presses were plotted, on the basis of the data obtained by interpreting the oscillograms, for the systems "accumulator plant and main plunger cylinder" and "accumulator plant - mandrel cylinder"; these characteristics are plotted in fig.26.

As a result of the static analysis of the experimental data, reduced coefficients of hydraulic resistance were found for the 1500 and 3000 ton presses: ξ_{gp} - 5.33, 9.75; ξ_{ig} - 3.30, 5.83.

Fig.25 General appearance of the apparatus for measuring tube temperature: (1) pyrometer head; (2) tube.

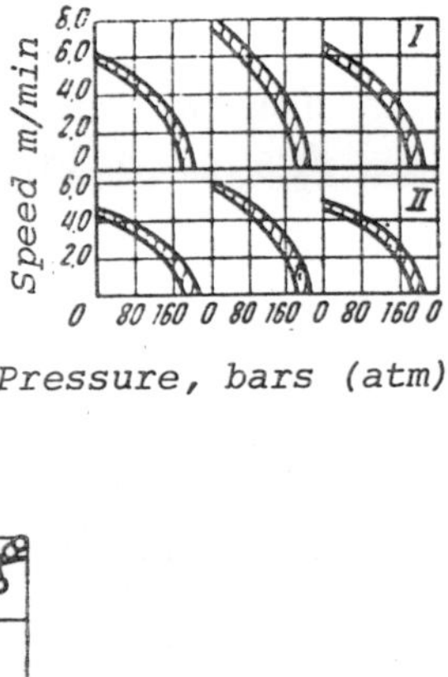

Fig.26 Speed characteristics for horizontal (I) 1500 tonne and (II) 3000 tonne presses: (a) system "accumulator plant - main ram cylinder"; (b) system "accumulator plant - mandrel cylinder"; (c) speed characteristic for the main plunger cylinder and mandrel cylinder working in conjunction.

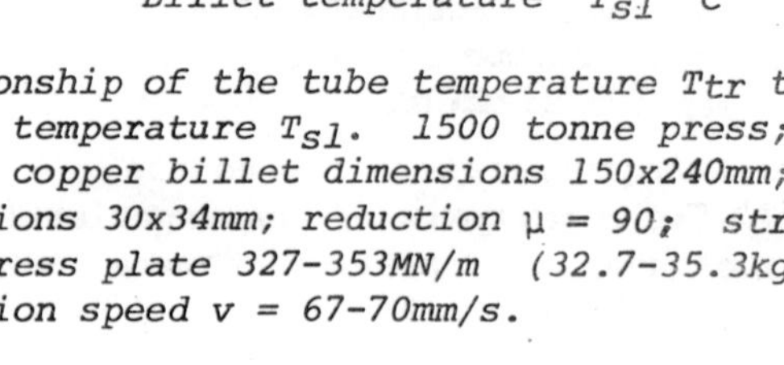

Fig.27 Relationship of the tube temperature T_{tr} to the billet temperature T_{sl}. 1500 tonne press; D_k = 155mm, copper billet dimensions 150x240mm; tube dimensions 30x34mm; reduction μ = 90; stress p in press plate 327-353MN/m (32.7-35.3kg/mm^2); extrusion speed v = 67-70mm/s.

During the extrusion of tubes, the mandrel plunger cross-member usually rests against the adjusting screw of the main ram. The extrusion force is therefore the sum of the forces P_{gp} and P_{ig} developed by the ram and mandrel cylinders respectively. In this case the relationship of the extrusion speed to the extrusion force can be determined as follows:

$$P_{gp} = (P_{akk} - \xi_{gp} v^2_{gp}) S_{gp} \quad (36)$$

$$P_{ig} = (P_{akk} - \xi_{ig} v^2_{ig}) S_{ig} \quad (37)$$

here S_{gp} and S_{ig} are the cross-sectional areas of the main plunger and mandrel cylinders respectively, in cm^2.

The total extrusion force is equal to:

$$P_{pr} = P_{gp} + P_{ig} = (P_{akk} - \xi_{gp} v^2_{gp}) S_{gp} + (P_{akk} - \xi_{ig} v^2_{ig}) S_{ig} \quad (38)$$

When the main plunger and the mandrel are acting together, $v_{gp} = v_{ig}$. If, then, we divide the right and left hand sides of equation (38) by $S_{gp} + S_{ig}$, we get:

$$p_{usl} = p_{akk} - \xi v^2_{gp} \quad (39)$$

here p_{usl} is the nominal extrusion pressure, equal to:

$$p_{usl} = \frac{P_{pr}}{S_{gp} + S_{ig}}$$

The coefficient of hydraulic resistance for the press:

$$\xi = \frac{\xi_{gp} S_{gp} + \xi_{ig} S_{ig}}{S_{gp} + S_{ig}} \quad (40)$$

The speed characteristics of 1500 and 3000 ton presses are given for this case in fig.26, for $\xi = 5.14$ and 9.28 resp. With constant fluid pressure, forces on the mandrel and on the head of the ram are directly proportional to the areas S_{gp} and S_{ig}. For the case under consideration the force on the mandrel constitutes 10% of that of the ram head for a 1500 ton press and 16.5% for a 3000 ton press. It follows from press speed characteristics (fig.26) that equality can only be reached between the main plunger and mandrel cylinder pressures with very low speeds, close to zero. If extrusion speed is varied within the range encountered in practice, the relationship between the main ram and mandrel cylinder forces alters as shown in table 10. In the general case the ratio p_{ig}/p_{gp} can be found using equations (36) and (37). It is possible, allowing for the part played by force developed by the mandrel ram in the total force with different extrusion speeds, to make fuller use of press power and accurately to determine temperature-speed characteristics.

Table 10

Variation in the extrusion speed according to the ratio of the main plunger and mandrel plunger forces

Press power MN (tons)	Extrusion speed m/min	Main plunger force MN (tons)	Mandrel cylinder force MN (tons)	Mandrel cylinder force as a percentage of the plunger force
1500	2.14-5.45	3.2-3.17 (320-317)	1.39-0.78 (139-78)	10.5-24.7
3000	1.2-3.7	22.8-8.2 (2280-820)	3.9-2.45 (390-245)	17.0-29.5

practical calculations, therefore, the extrusion force is usually taken as being 75-85% of the nominal press power (16,5,3).

In the case of 1500 ton presses, the extrusion force is 11.3-12.8 MN (1130-1280 tons). The pressure of the liquid in the main ram will in this case be 15-16.9MN/m^2 (150-169atm), and according to the diagram in fig.26 the extrusion speed will be 3.04-2.41m/min. With these extrusion speeds the force developed by the mandrel ram is 1.265-1.355MN (126.5-135.5 tons). When the mandrel and main ram cylinders are working together the force is 12.565-14.15MN (1256.5-1415.0 tons), and these figures amount to 82-94.0% of the nominal press pressure.

When the main plunger and the mandrel ram are working together, therefore, the permissible press pressure for practical calculations can be recommended as 85-95% of the rated pressure. It should be noted that, if extrusion is calculated solely on the basis of the force developed by the main cylinder, the combined running of the mandrel and main ram cylinders will increase the speed. For instance, with a prescribed extrusion force of 1280 tons in an 1500 ton press, with the main ram alone working the extrusion speed will be 2.41m/min (fig.29). With the same extrusion force but the main ram and the mandrel cylinder working in conjunction, the extrusion speed will be equal to 2.99m/min, i.e. 20% greater than in the first case. The increase in the extrusion speed is due to the fact that, in the first case, load is taken from the main ram and the extrusion speed increases accordingly.

Most industrial alloys are extruded with limited metal flow speeds owing to the danger of disturbing the continuity of the metal. When the production parameters (the dimensions of the billet and the finished tube, the withdrawal coefficient and the temperature) are calculated however, the extrusion speed is not determined, owing to the fact that high-speed characteristics for presses are not available.

The operator has to vary the speed of flow directly during the running of the plant. If cracks develop in the surfaces of bars or tubes, the operator does not fully open the filling valve, or, though this is much less frequently done, he partly closes the throttling valve. This increases the coefficient of hydraulic resistance ξ and reduces the rate of metal flow. This coarse adjustment of the extrusion speed does not, however, always prevent cracking.

The press speed characteristic enables us accurately to calculate the rate of flow for any particular case of extrusion, and to alter it, if it is excessive, by altering the reduction or lowering the temperature to which the metal is heated.

The following equation is true for running idle, when $p_{us} = 0$ with the main plunger and the mandrel moving together ($v_{gp} = v_{xx}$):

$$0 = p_{akk} - \xi v^2_{xx} \qquad (41)$$

If we transfer p_{akk} in equations (39) and (41) from the right to the left hand side, and then divide the right and left hand sides of equations (39) and (41) by one another, we get:

$$\frac{p_{us} - p_{akk}}{-p_{akk}} = \frac{v^2_{gp}}{v^2_{xx}}$$

whence

$$v_{gp} = v_{xx} \sqrt{1 - \frac{p_{us}}{p_{akk}}} \qquad (42)$$

When only the main plunger is operating, in an analogous manner we obtain the following equation from equations (34) and (41):

$$v_{gp} = v_{xx} \sqrt{1 - \frac{p_{gp}}{p_{akk}}} \qquad (43)$$

If, therefore, we know the extruding force p_{gp} and the idle running speed v_{xx} we can determine the extrusion speed v_{gp}. It should, however, be taken into account that the value of the coefficient ξ for idle running may not be the same as for extrusion, owing to the different running of the packing glands and the lower frictional losses in the cylinder.

Temperature conditions for extrusion

During the hot extrusion of non-ferrous metal and alloy tubes, the temperature of the metal is of prime importance, since the quality of the product, and the power consumption on deformation, i.e. the power of the press, depend on it. The temperature to which initially to heat billets is selected on the basis of the equilibrium diagram and mechanical properties of the metal, which govern its ductility. The metal does not however remain at a constant temperature during extrusion, and its temperature depends on that of the billet. The reason for this is that two contradictory phenomena take place during the extrusion process: the billet is cooled by conduction and the metal is heated by heat equivalent to the work of deformation.

The metal cools during extrusion, as a result of conduction through the outer and inner (when tubes are extruded) surfaces of the billet. In general the amount of heat transferred into the pool around the billet depends on the extrusion speed, the temperatures of the billet and the press tools, their thermal conductivities and heat capacities, and the properties of the contacting surfaces and the coating applied to them.

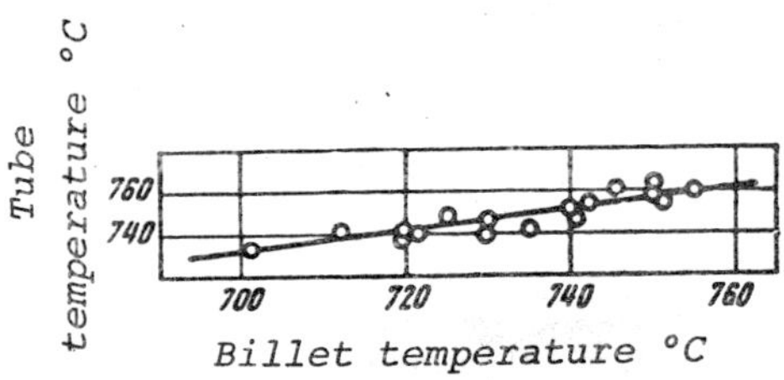

Fig.28 Relationship of tube temperature to billet temperature. 1500 tonne press, D_k = 155mm, L62 brass billet, tube dimensions 35x39mm; μ = 77; p = 461-504MN/m^2 (46.1-50.4kg/mm^2); v = 49-62mm/s.

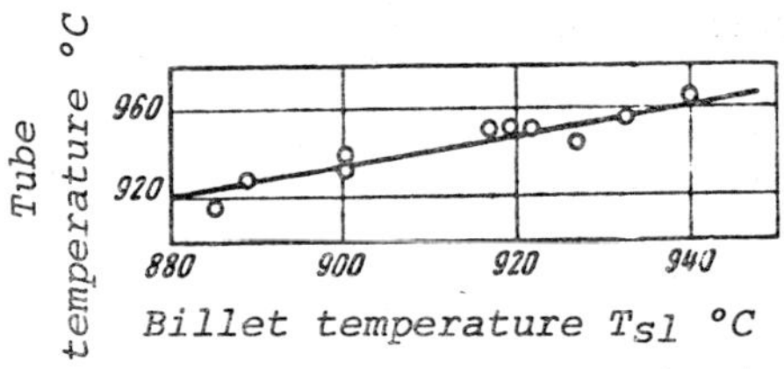

Fig.29 Relationship of the tube temperature to billet temperature. 1500 tonne press, D_k = 155mm, MNZh5-1 alloy billet, tube dimensions 35x4mm; μ = 50; p = 618-662MN/m^2 (61.8-66.2kg/mm^2); v = 38-42mm/s.

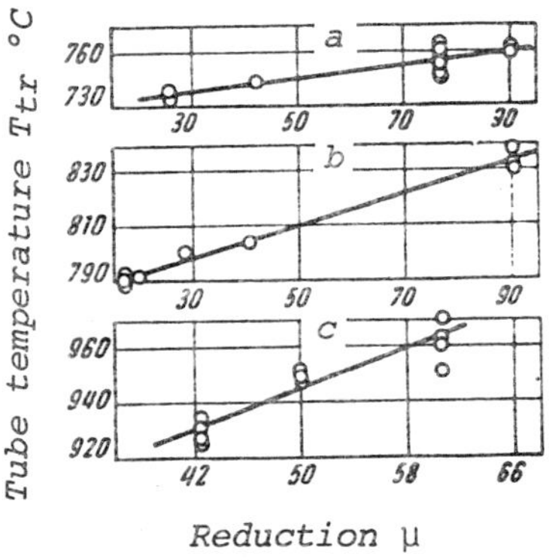

Fig.30 Relationship of tube temperature T_{tr} to the reduction μ with T_{sl} ≈ const. 1500 tonne press, D_k = 155mm. Ingot dimensions 150x240mm: (a) L62 brass; (b) copper; (c) MNZh5-1 alloy.

The degree of deformation, the extrusion speed and pressure, the temperature of the billet, and other parameters such as the temperature of the container etc., govern the differences between the temperatures of the tube and the billet, and the alteration in the tube temperature during extrusion. Owing to the large number of variable factors, it is not possible theoretically to determine the temperature of the metal leaving the die and the change in its temperature during extrusion, and the temperature of the metal in the press tools can only be gauged on the basis of experimental data for each particular case.

It is particularly necessary to allow for temperature changes when extruding tubes made of alloys with narrow extrusion temperature ranges. When these alloys are extruded, to ensure that their quality is good along their length it is necessary to maintain the temperature of the metal leaving the die within very definite limits, i.e. to ensure isothermal extrusion process conditions. Of all the methods of regulating temperatures, practically the only possible method is to alter the extrusion speed, while simultaneously controlling the temperature of the metal leaving the die by means of a partial radiation or spectral ratio optical pyrometer. Most of the experiments which have been performed have, however, related to processes in which cooling of the billet in the container has been practically eliminated by preheating the latter. In this case heat equivalent to the work of deformation as a rule entirely compensates the loss of heat due to cooling, and the temperature of the tube often rises by the end of extrusion (48). In most cases it is undesirable that the temperature should vary along the tube, since this causes inconsistent mechanical properties and dimensions along the tube, also red shortness in a number of alloys with narrow ductility temperature ranges, etc.

Heavy non-ferrous metals and alloys, also steels, titanium and other metals, are extruded in containers the temperature of which is much lower than that of the billet. For this reason study of the extrusion temperature conditions and their relationship to technological parameters is of definite interest in connection with accelerating the process.

If we assume the distribution of sources of heat in a billet to be as stated by Singer in reference (48), and also assume that the heat from the forces of friction at the contact surfaces of the container and the mandrel is consumed on heating them, the extent to which the temperature of the metal is raised owing to the work of deformation can be determined using the equation:

$$\Delta T_{def} = k p_{pr} \qquad (44)$$

where k is the coefficient of rise in temperature, equal to:

$$k = \frac{1}{1.8c \cdot \gamma} \cdot 10^{-8}$$

or:

$$k = \frac{1}{42\ 700c\gamma}$$

here:

- c = the specific heat capacity of the metal being extruded within the range of temperatures under consideration in J/kg·°C (kcal/g·°C);
- γ = the density of the metal in kg/m^3;
- p_{pr} = the specific extrusion pressure (the stress in the pressure pad) for a length of billet equal to the length of the region of deformation in N/mm^2 (kg/mm^2) $p_{pr} = p_{min}$.

If the temperature T_{sl} of the billet, and the temperature T_{tr} of the tube on leaving the die are known, also the value of ΔT_{def} from equation (44), the absolute value of the degree of cooling $\Delta T_{zakh.a}$ can be determined using the following equation:

$$\Delta_{zakh.a} = T_{sl} + \Delta T_{def} - T_{tr} \qquad (45)$$

The calculation results are given in Table 11. In every case the time between the start of extruding a billet and when the tube temperature is measured is 2.2-2.6s. The cooling of the billet in the container during extrusion, piercing and the first stage of extrusion, is very important, and is basically governed by the thermal conductivity of the alloy concerned. For instance in the case of copper the cooling of the metal, allowing for the heat from the work of deformation, is 193-199°C. The thermal conductivity of the MNZh5-1 alloys is slightly lower than that of copper, and the cooling amounts to 146-154°C. The thermal conductivity of L62 brass is even lower, and in this case the cooling amounts to 102-109°C. The data obtained indicate that, in cases in which the temperature of the container is lower than that of the billet, the heat from the work of deformation mostly does not heat the metal, but is transferred into the colder tools (container, mandrel, pressure pad and die).

The true cooling of the metal $\Delta T_{zakh.f}$ is more important than the absolute cooling $\Delta T_{zakh.a}$ as regards technological calculations and the particulars of the metal after extrusion; the true cooling is equal to the difference between the temperature T_{tr} of the tube metal and the initial temperature T_{sl} of the billet, i.e.

$$\Delta T_{zakh.f} = T_{sl} - T_{tr} \qquad (46)$$

Table 11

Cooling of the metal during the extrusion of tubes in a 1500 ton press. Container diameter D_k = 155mm, Billet length L_{sl} = 240mm. The time τ between the start of extrusion and temperature measurement is 2.2-2.6s with a billet length $L_x \approx$ 210 mm

Metal	Tube Dimensions $d_{ig} x D_{tr}$ mm	Reduction μ	Stress p in press plate MN/m^2 (kg/mm^2)	T_{sl} °C	T_{tr} °C	$\Delta T_{zakh.f}$ by equation (46)	k mm^2•°C/kg	T_{def} °C	$\Delta T_{zakh.a}$ by equation (45)
Copper	30 × 34	90.0	590-610(59-61)*	870-880	830-840	40	2.59	153-160	193-199
MNZh5-1	55 × 63	31.2	470-520(47-52)	925-935	900-915	20-25	2.42	116-129	146-154
L62	35 × 39	77.0	440-460(44.46)	740-750	745-760	5-10	2.5	112-114	102-109

* Extrusion was carried out after extruding L62 alloy

The value of $\Delta T_{zakh.f}$ may fluctuate very widely, and is governed by the temperature to which the billet is initially heated, the extrusion speed, the alteration in this during the plunger travel, the reduction μ, etc. Figures 27-29 show how the temperature of the metal leaving the die alters with relation to the temperature to which the billet is preheated.

The increase in the billet preheat temperature disproportionately raises the temperature of the metal on leaving the die. The reason for this is that, as the billet temperature is raised, the extrusion pressure decreases and consequently also the heat from the work of deformation. The amount of heat lost by the billet will be greater owing to the increased temperature gradient between the billet and the tool. Assuming that the extruding pressure p_{pr} is directly proportional to the yield stress σ_s, we can determine the reduction in this heat from the work of deformation for particular cases:

$$\frac{\Delta T^n_{def}}{\Delta T^v_{def}} = \frac{p^n_{pr}}{p^v_{pr}} = \frac{\sigma^n_s}{\sigma^v_s} \qquad (47)$$

We can see from the data given in Table 12 that the heat from the work of deformation was reduced by 1.18% for copper, by 1.58% for L62 brass and by 1.21% for the MNZh5-1 alloy. The value of ΔT_{def} dropped by 23, 46 and 33°C respectively.

Increasing the temperature of the billet results in increase in the true cooling $\Delta T_{zakh.f}$. We can see from the figures in Table 13 that, while at the lower temperature limit (T^n_{sl} = 870°C) $\Delta T_{zakh.f}$ was equal to 50°C for copper, at the upper limit (T^v_{sl} = 930°C) the true cooling was 60°C. Tubes made of L62 brass and MNZh5-1 alloy were at higher temperatures than the billets (minus sign), but as T_{sl} was increased this difference decreased, i.e. more cooling took place.

The difference in the true cooling figures for the MNZh5-1 alloy, L62 brass and copper (see Table 13) is due firstly to the fact that the thermal conductivity of copper is much higher than that of L62 brass and of MNZh5-1 alloy. Secondly it follows from equation (44) that the rise in the temperature in the region of deformation is proportional to the specific extrusion pressure p_{pr} and the coefficient k of rise in temperature.

We can see from the values given in Table 12 that the coefficient k of rise in temperature varies little, while the specific extruding pressure for the MNZh5-1 alloy tubes under consideration is respectively 44% and 28% higher than for the extrusion of L62

Table 12

Effects of billet temperature T_{sl} on tube temperature T_{tr} during the extrusion of tubes in a 1500 ton press; D_k = 155mm, T_{sl} = 240mm. Billet length L_x ~210mm

	Tube dimensions $d_{ig} \times D_{tr}$ mm	Reduction μ	k	Ingot heating temperature					
				Lower					
				T_{sl} °C	σ_s^n MN/m² (kg/mm²)	p MN/m² (kg/mm²)	ΔT_{def} °C	T_{sl} °C	σ_s^v MN/m² (kg/mm²)
Copper	30 x 34	90	2.59	870	19.5(1.95)	590(59)* 350(35)	153 91	930	16.5(1.65)
L62	35 x 39	77	2.42	700	15.0(1.50)	520(52)	126	760	15.8(1.58)
MNZh5-1	35 x 41	50	2.50	880	46.0(4.60)	750(75)	188	940	12.1(1.21)

* Extrusion was carried out after extruding L62 alloy
† When T_{sl} is increased by 60°C

limits				Change in tube temperature T_{tr}†		Time after start of extrusion to tube temperature measurement τ in s
Upper						
σ_s^n/σ_s^v	p_{calc} MN/m^2 (kg/mm^2)	p_{true} MN/m^2 (kg/mm^2)	ΔT_{def}^v °C	Calculated $\Delta T^n_{def}-\Delta T^v_{def}$	True $T^n_{tr}-T^v_{tr}$	
1.18	500(50)*	470(47)	130	23	50	2.4-3.0
	300(30)	-	78			
1.58	330(33)	-	80	46	30	2.1-2.4
1.21	620(62)	610(61)	155	33	40	2.2-2.7

brass and copper tubes. The heat lost in the extrusion of copper alloys is not compensated by heat from the work of deformation, and the temperature of the metal on leaving the die is lower than the initial temperature of the billet. In the case of the MNZh5-1 alloy the metal is heated more owing to heat from the work of deformation of the billet and the temperature of the metal on leaving the die is higher than the initial temperature of the billet. The L62 brass occupies an intermediate position.

Table 13

Effects of billet temperature T_{sl} on the true cooling of the metal $T_{zakh.f}$ according to equation (46) for the extrusion of tubes in a 1500 ton press, D_k = 155mm, L_x = 210mm

Metal	Tube dimensions mm	Billet temperature T_{sl} °C	True cooling $T_{zakh.f}$ °C
Copper	30 x 34	870	50
		930	60
L62	35 x 39	700	-20
		760	0.0
MNZh5-1	35 x 41	880	-40
		940	-20

As the reduction μ is increased the temperature of the metal in the region of deformation rises, and, as the experimental data show, increase in the reduction μ also raises the temperature of the tube (fig.30). The theoretical rise in tube temperature when the reduction is increased from μ' to μ'' can be calculated using the equation:

$$\Delta T_{tr} = \Delta T''_{def} - \Delta T'_{def} = k(p''_{pr} - p'_{pr}) \qquad (48)$$

here p'_{pr} and p''_{pr} are the respective stresses in the pressure pad with reductions of μ' and μ''. Comparison of the calculated and true values of the rise in temperature (Table 14) shows that they agree well.

During extrusion, the temperature of the metal on leaving the die does not remain constant, but alters during the travel of the main plunger. Figures 31 and 32 show how the temperature of a finished tube leaving the die varies, also how the extrusion speed and extrusion pressure vary along a billet. In every case, tube temperature drops during extrusion. The alteration in tube temperature between the start and end of extrusion depends on the extrusion speed and pressure. With low extrusion speeds, the

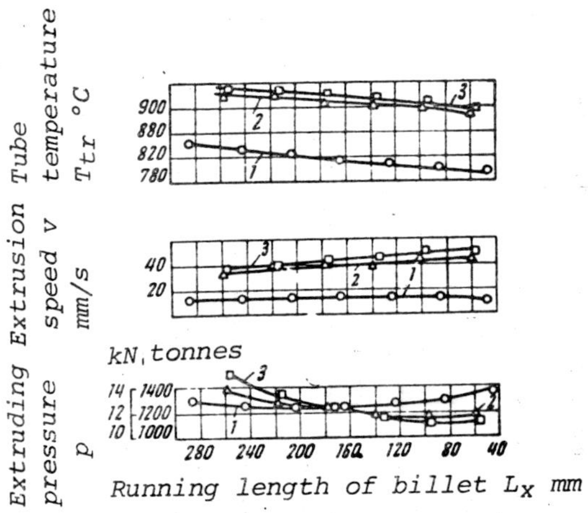

Fig.31 Variation in tube temperature T_{tr}, the pressure p on the pressure pad and the extruding speed v, during the extrusion of tubes in a 1500 tonne press. MNZh5-1 alloy. 175x300mm billet, tube dimensions 55x63mm, reduction μ = 31.2, billet temperature 930°C. Initial extruding speeds: (1) 12mm/s (2) 30mm/s (3) 36mm/s.

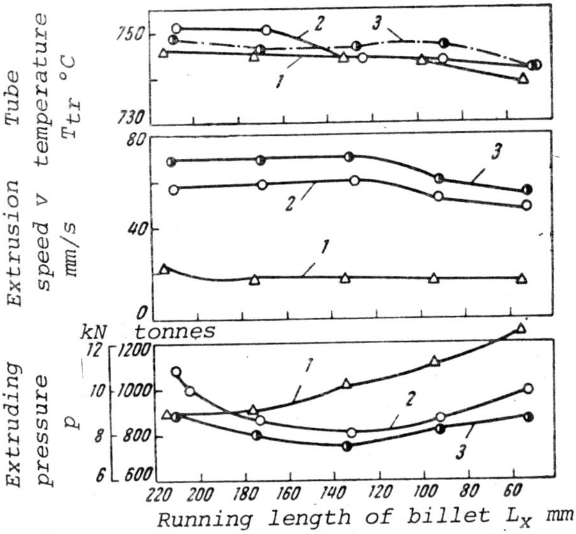

Fig.32 Variation in tube temperature T_{tr}, the pressure p on the pressure pad and the extrusion speed v, during the extrusion of tubes in a 1500 tonne press. L62 brass. 150x240mm billets; tube dimensions 35x39mm, reduction μ = 77; billet temperature 750°C. (1) initial extrusion speed 22mm/s; (2) 58mm/s; (3) 72mm/s.

Table 14

Variation in tube temperature as the reduction is increased during the extrusion of tubes in a 1500 ton (15MN) press. Container diameter D_k = 155-180mm

Metal	Tube dimensions $d_{ig} \times D_{tr}$ mm	Billet temperature T_{s1} °C	Reduction μ	Stress in pressure pad p (MN/m²) kg/mm²	Rise in tube temperature ΔT_{tr}	
					Calculated with equation (44)	Experimental
Copper	30 x 50	835-870	14.3	(375-392) 37.5-39.2	-	-
	30 x 34	860-880	90.0	(587-592) 58.7-59.2	50-57	50
L62 brass	30 x 42	710-720	26.7	(442-458) 44.2-45.8	-	-
	30 x 34	710-720	90.0	(615-627) 61.5-62.7	39-46	30
MNZh5-1 alloy	55 x 63*	910-930	31.2	(489-497) 48.9-49.7	-	
	35 x 40	910-940	60.6	(709-720) 70.9-72.0	52-56	40

*Container diameter D_k = 180mm

pressure increases by the end of extrusion because the billet has cooled considerably. The extrusion speed has little effect on the temperature of L62 brass tubes (fig. 32).

Figure 33 is an equivalent diagram for the change in the temperature of copper tubes on leaving the die and the extrusion speed and pressure, with different reductions. With the same extrusion speeds ($v \approx$ 60mm/s) and initial billet temperatures (T_{sl} = 850 C), the higher the extrusion pressure the higher the temperature of the tube (T_{tr}) when it leaves the die.

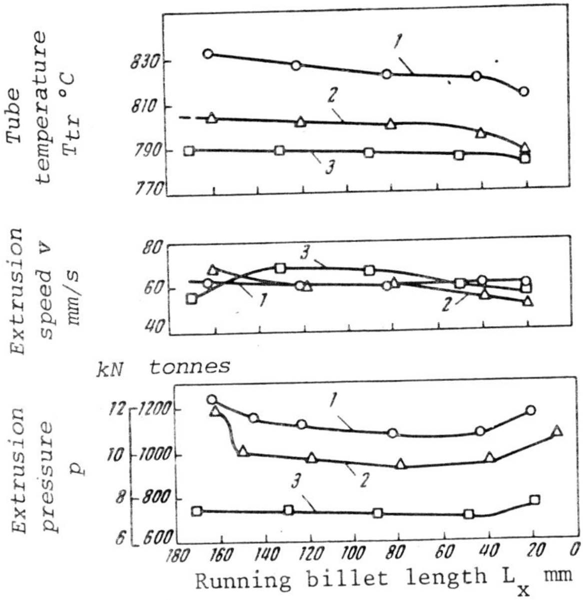

Fig.33

Variation in the tube temperature T_{tr}, the extrusion speed v and the extrusion pressure P', during the extrusion of tubes in a 1500 tonne press with different reductions. Copper. 150x240mm billet; billet temperature 850°C; (1) tubes dimensions 30x34mm, μ = 90; (2) tube dimensions 30x38mm, μ = 41.3; (3) tube dimensions 30x50mm, μ = 14.5

The basic factor affecting changes in the temperature of the metal during extrusion is therefore the extrusion speed. As a result of our research, we obtained relationships of the temperature of tubes on leaving the die to the extrusion speed at any moment of time (with a running billet length L_x) (fig.34). This enables us to plot the relationship of variation in the extrusion speed, as the main plunger travels, with which the temperature of the tube on leaving the die remains constant.

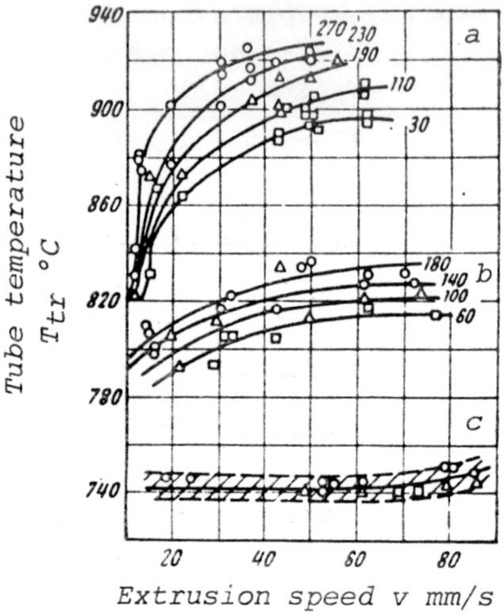

Fig.34

Relationship of the tube temperature T_{tr} to the extrusion speed. The figures against the curves denote the running billet length L_x in mm (a) MNZh5-1 alloy, 175x300mm billets, tube dimensions 55x63mm, reduction 31.2 billet temperature 920-930°C; (b) copper 150x240mm billets, tube dimensions 30x34mm, reduction μ = 90; billet temperature 860-880°C; (c) L62 brass 150x248mm billets, tube dimensions 35x39mm, reduction 66 billet temperature 750x760°C.

Figure 35 shows how the extrusion speed varies for the MNZh5-1 alloy.

As opposed to the results obtained by Singer (48) and Laue (50), when aluminium or copper are extruded in a container the temperature of which is equal to that of the billet, in order to provide isothermal extrusion conditions for MNZh5-1 alloy tubes the extrusion speed is increased before the end of the operation (fig.35)

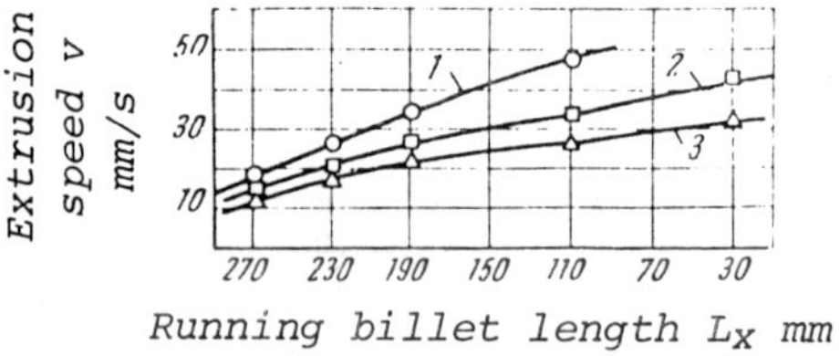

Fig.35

Variation in the extrusion speed along the main plunger with isothermal extrusion conditions. MNZh5-1 alloy 175x300mm billet, tube dimensions 55x63mm, reduction μ = 31.2 with the following different tube temperature (1) 900°C, (2) 890°C, (3) 880°C.

Study of the temperature conditions for extrusion makes it possible to use automatic systems for regulating the extrusion speed in order to provide isothermal conditions for the process (fig.36).

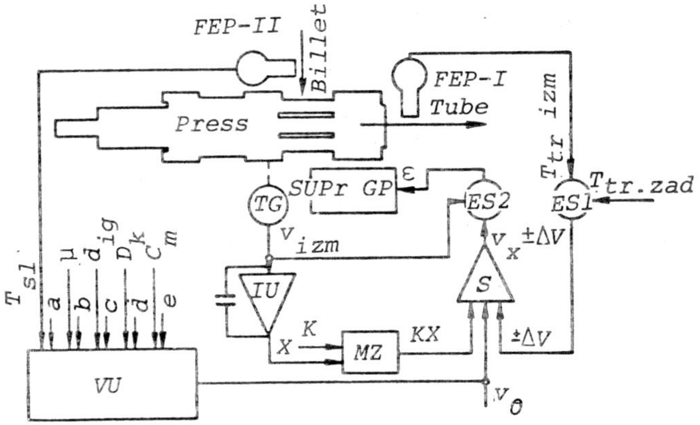

Fig.36 Diagram of automatic regulation of the extrusion process to provide isothermal conditions: (VU) calculating device; (MZ) multiplying link; (S) summing unit; (IU) integrating device; (ES) comparison element; (SUPr GP) system for controlling the main plunger drive; (FEP) photoelectric pyrometers; (TG) tachogenerator.

In order to achieve accurate regulation, the system includes a feedback on the temperature; this feedback system is as follows. When a tube leaves the die, the photoelectric pyrometer FEP-1 measures it temperature. The voltage taken from FEP-1 ($T_{tr.izm}$) is compared with a master voltage proportional to the tube temperature required according to the technology ($T_{tr.zad}$). If the temperature of a tube leaving the die is higher or lower than the prescribed temperature, a voltage ΔV appears at the output from the comparison element ES1; this voltage is proportional to the temperature difference, and may be plus or minus. This voltage $\pm\Delta V$ is fed to the summator S, which carries out correlation of speed to temperature: depending on its sign, ΔV is added to or deducted from the voltage v_x proportional to the calculated extrusion speed.

If, therefore, the temperature of a tube leaving the die is above the prescribed temperature, a mismatch ΔV immediately appears and this reduces the calculated extrusion speed. On the contrary, if the tube temperature is lower than the prescibed temperature

(+ ΔV), the extrusion speed is increased and the temperature of the tube rises.

The voltage $v_x \pm \Delta V$ obtained in the comparison unit ES2 is compared with the voltage v_{izm} proportional to the true extrusion speed which is taken from the tachogenerator TG connected to the moving cross member of the main plunger. If these voltages are not equal, i.e. if at any particular moment the extrusion speed does not correspond to the prescribed value of $v_x \pm \Delta V$, a mismatch ε with a particular sign appears at the output from ES2. This mismatch will act on the system controlling the main plunger drive (SUPrGP) opening or closing the throttling slide valve.

In order to obtain the calculated extrusion speed (the voltage v_x proportional to this speed), the parameters of an equation representing the change in speed with relation to the prescribed tube temperature and the billet temperature T_{sl} measured with the pyrometer FEP-II, the reduction μ, the container diameter D_k, the mandrel diameter d_{ig}, the thermal conductivity λ of the alloy, and the constants (a,b,c,d,e) which are found by experiments, are fed to the special calculating device VU.

The initial value of the speed v_0 is obtained at the output from VU. This speed is established before extruding a billet $(x = 0)$. As soon as the main plunger has begun to move at a specific speed a voltage proportional to the true speed, v_{izm}, is taken from the tachogenerator TG. This voltage passes through the integrating device IU and is transformed into a voltage "x" proportional to the main plunger travel. The voltage x is fed to the multiplying link MZ, into which the experimental coefficient k is also fed; this coefficient allows for the necessary increase in speed along the plunger travel. The product thus obtained is added, in the summator S, to the initial prescribed speed v_0.

The result is the calculated extrusion speed (voltage) equal to:

$$v_x = v_0 + kx \qquad (49)$$

The feedbacks for speed and temperature make the system stable and eliminate all the inaccuracies which may have been permitted in the calculations. If there is no calculating device the voltage corresponding to v_0 can be found by experiments.

EXTRUSION FORCE

When tubes are extruded using cylindrical mandrels the total extruding pressure (the pressure on the pressure pad) is made up of the forces developed by the main plunger and mandrel cylinders. The exception to this is extrusion with bottle-shaped mandrels, whe[n] the moving cross-member of the mandrel plunger rests against supporting adjusting nuts and the force on the pressure pad is created by the main plunger alone.

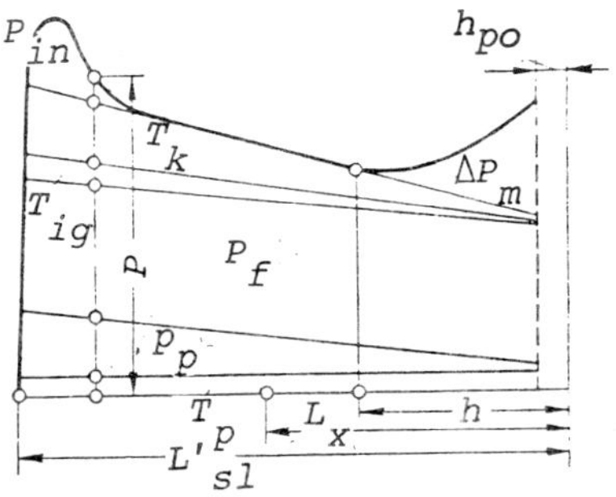

Fig.37 Variation in the extruding pressure p along a billet

Figure 37 is a standard diagram for the variation in the extruding force along a billet (the plunger travel) for the extrusion of non-ferrous metal tubes by the direct method. The extrusion pressure is made up of the following principal components:

1) the force expended on actually changing the shape of the metal in the squeezing zone of the region of deformation - P_f;

2) the force expended on displacing particles of metal within the billet outside the region of deformation - P_p;

3) the forces associated with overcoming friction along the contact surfaces of the container and the mandrel - T_k and T_{ig} respectively;

4) the force expended on overcoming friction in passing through the sizing zone of the die - T_m;

5) the forces associated with overcoming billet inertia, the transition from contact friction of rest to sliding friction, the thermal effect of deformation, and certain other phenomena (4) - P_{in};

6) the increase in the deformation forces at the end of the extrusion process associated with increase in the forces of friction around the pressure pad when this enters the squeezing region of the plastic zone and deformation of the metal in the "dead" (elastic) zone - ΔP_m.

The total extruding pressure is thus the sum of the components:

$$P = P_f + P_p + T_k + T_{ig} + T_m + P_{in} + \Delta P_m \quad (50)$$

The component P_p depends on the length of the billet, as was proved in reference (22), and the nature of metal flow. With laminar flow in the metal being extruded $P_p \approx 0$. In the case of turbulent flow (4), P_p decreases as billet length decreases, and with a billet length equal to that of the region of deformation $P_p = 0$. In practical calculations the magnitude of P_p is often not taken into account (4), since it is relatively low by comparison with the total extruding pressure. In most cases the force P_f does not remain constant throughout the plunger travel; owing to the metal cooling, the strength characteristics of the metal being extruded usually rise by the end of extrusion, and this results in an increase in P_f.

It was proved in references (5, 51-53 etc.) that I L Perlin's theoretical equation (54,55) provides extremely accurate results when used for calculating the extruding force. Since, however, the equation is cumbersome and there are no available data as to the forces of friction at the contact surfaces, engineering calculations using this equation are made more difficult. Inaccurate determination of the frictional stress (in I L Perlin's equation it is necessary to determine three values of the frictional stress) results in considerable errors when the pressure is calculated.

I L Perlin (4) defined the force P_f associated with change in the shape of the metal as the sum of two components; the force R_m for overcoming internal friction in the region of deformation, and the force T_m required to overcome contact friction along the mandrel and along the inner surface of the dead zone:

$$P_f = R_m + T_m$$

while:

$$R_m = \beta \frac{\pi}{4} \left(\frac{D_k^2}{\cos^2 \frac{\alpha}{2}} - \frac{d_{ig}^2}{\cos^2 \frac{\phi}{2}} \right) \sigma_{sr} \ln \mu$$

and

$$T_m = \frac{\pi}{2 \sin \alpha} (D_k - d_{ig}) \tau_{sr} \ln \frac{D_k - d_{ig}}{D_{tr} - d_{ig}}$$

where:

β = the coefficient representing the effect of the mean normal main stress; for extrusion, $\beta \approx 1.1$;

α = the angle of taper of the squeezing zone or the die; with a plane die, or with a die taper angle $\alpha \leqslant 60°$, for the calculation we can take it that $\alpha = 60°$:

$$\phi = \arcsin \left(\frac{d_{ig}}{D_k} \cdot \sin \alpha \right);$$

σ_{sr} = the mean yield stress in the plastic zone;

τ_{sr} = the mean plastic shear stress in the metal being extruded, at the side surface, in the squeezing part of the plastic zone;

$ln\mu$ = the integral index of the degree of deformation;

d_{ig} and D_k = diameters of the mandrel and the container respectively.

Assuming that $\tau_{sr} \approx 0.5\ \sigma_{sr}$ it is easy to prove that the stress P_f in the pressure pad associated with the change in the shape of the metal is equal to:

$$P_f = A\sigma_{sr}\ ln\ \mu \qquad (51)$$

where:

$$A = \frac{\beta}{D_k - d_{ig}}\left(\frac{D_k^2}{\cos^2\frac{\alpha}{2}} - \frac{d_{ig}^2}{\cos^2\frac{\phi}{2}}\right) + \frac{1}{\sin\alpha} \cdot \frac{1}{ln\ \mu} \cdot ln\ \frac{D_k - d_{ig}}{D_{tr} - d_{ig}}$$

depends on the parameters of the region of deformation ($D_k\ d_{ig}\ \alpha\ \phi$) and the nature of the stressed state (β).

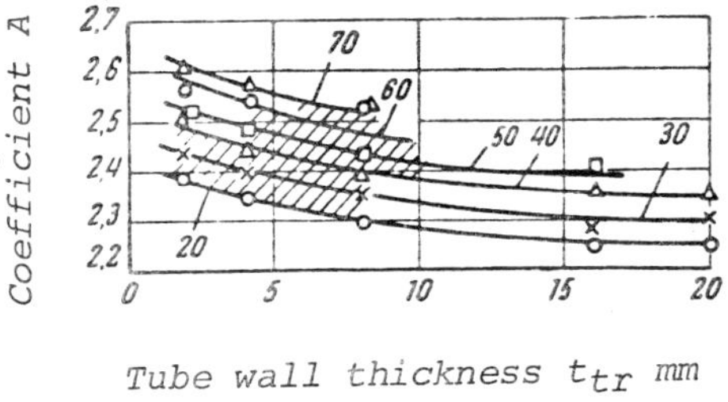

Fig.38 Relationship of the coefficient A to tube wall thickness t_{tr}. The figures above the curves denote the diameters of tubes extruded. Container diameter D_k = 180mm.

Figure 38 shows how the coefficient A varies, with $\alpha = 60°$, for a wide range of tubes (the figures against the curves are extruded tube diameters). The coefficient A varies between 2.3 and 2.7 for the range of tubes extruded in practice in 1500 ton and 3000 ton presses (shown shaded in fig.39). If we take A = 2.5, the error in making engineering calculations of the force P_f using equation (51) does not exceed ±8%.

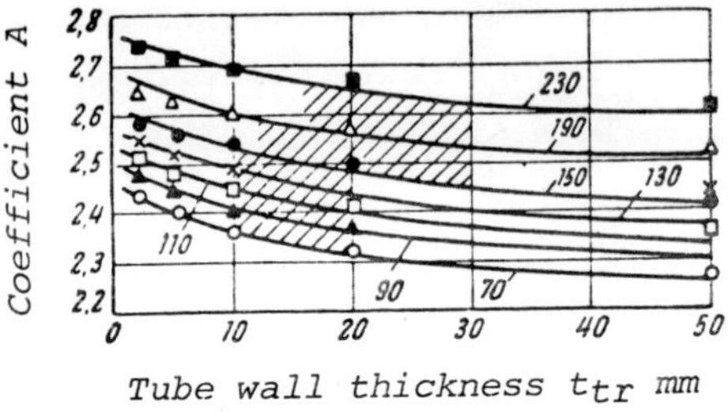

Fig.39 As for fig.38. Container diameter D_k = 408mm.

Table 15

Length of the region of deformation during the extrusion of copper tubes in a 1500 tonne press

Tube dimensions mm	Container diameter D_k mm	Length of back-extruded ingot L_{s1} mm	Length h of region of deformation mm	
			By equation (52) with $\alpha = 60°$	From oscillograms
26 x 34	180	210-220	93	100-110
30 x 42	180	220-250	91	100-110
45 x 51	180	240-250	87	95-110
60 x 66	180	220-250	81	100-105
26 x 33	155	250-260	78	110-115
30 x 37	155	250-260	77	95-115

Table 15 contains values of the length h of the region of deformation, found from oscillograms and by calculating. On the basis of the concept of the region of deformation as a spherical truncated segment, the length h of the region of deformation can be determined using the equation:

$$h = \ell_{o.d} = \frac{1}{2 \sin}\left[D_k \cos \phi - D_{tr} \cos \alpha\right] \qquad (52)$$

The values of the length of the region of deformation found using equation (52) are lower than those found from oscillograms. The reason for this is that the extruding pressure is increased as the metal cools and the minimum on the pressure curve shifts away from the die. With ratios $\frac{d_{ig}}{D_k} < 0.4$ the length of the region of deformation can be determined accurately enough using equation (24), since in this case cos $\phi \sim 1$.

With the minimum pressure, the total extruding force P_{min} is made up of the following components:

$$P_{min} = P_f + T_m$$

The relationship of $\frac{p_{min}}{\sigma_s}$ to the integral index of deformation ln μ for copper-base alloys can be represented by a straight line equation with no free term (fig.40). (p_{min} was found by dividing P_{min} found from the oscillograms by the press plate area $\frac{\pi}{4}(D_k^2-d_{ig}^2)$. This indicates that the component T_m has little effect on the extruding force. The relatively small importance of T_m was confirmed by calculation in reference (53); these calculations showed that this force is not more than 3-5% of the total extruding force.

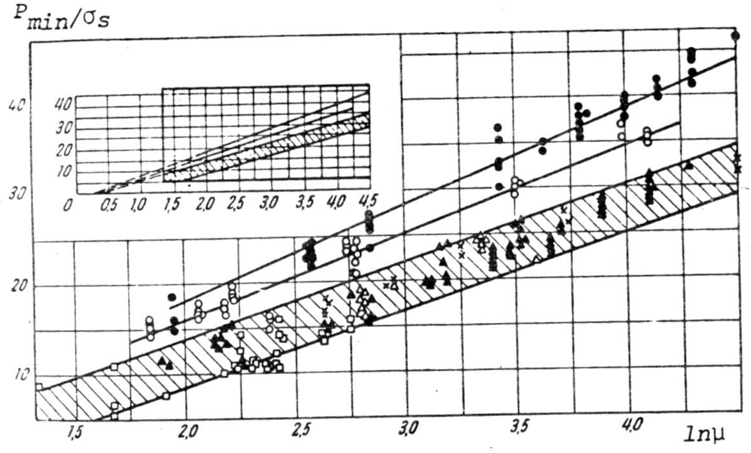

Fig.40 Relationship of the ratio of the minimum stress p_{min} in the press plate to the yield stress σ_s of the metal being extruded to the integral index of deformation ln . 1500 and 3000 tonne presses. Metals (●) L62 brass; (○) LS59-1; (X) copper; (▲) MNZh5-1; (△) MNZhMts30-0.8-1.0; (■) BrANZh

In making practical calculations it can thus be assumed, with some accuracy that:

$$p_{min} = p_f = A\sigma_{sr} \ln \mu \tag{53}$$

The strength properties of the metal being extruded are represented in equation (53) by the mean yield stress σ_{sr}, which depends on the temperature, the extrusion speed (or time) and the degree of deformation. It can be assumed, making certain assumptions (4), that on entering the region of deformation the metal has a yield stress σ_{sn} equal to the yield stress σ_s found by static compression or tensile tests in a linear stressed state. After deformation, on leaving the die the metal has a yield stress equal to:

$$\sigma_{sk} = k\sigma_{sn} = k\sigma_s$$

here k is the coefficient of strengthening.

It was proved in references (4,56) that, with the high deformations inherent in extrusion, it is best to determine the mean yield stress by geometrical meaning, i.e.

$$\sigma_{sr} = \sqrt{\sigma_{sn}\sigma_{sk}} = \sqrt{\sigma_{sn}k\sigma_{sn}} = \sigma_s\sqrt{k} \tag{54}$$

The coefficient of strengthening k is greatly affected by the weakening associated with the length of time spent by the metal in the plastic zone, also by the degree of deformation and the temperature changes in the extrusion conditions.

The deformation time τ_{def} during extrusion is determined by the equation, given in reference (4):

$$\tau_{def} = \frac{V_{pz}}{V_{sek}} = \frac{V_{pz}}{V_{p.sh} \cdot F_{sl}} \tag{55}$$

here:

V_{pz} = the volume of the squeezing part of the plastic zone;

V_{sek} = the volume of metal per second;

$V_{p.sh}$ = the extrusion speed (the speed of movement of the press plate;

F_{sl} = the transverse cross-sectional area of the pierced billet.

We can see from equation (55) that, as the extruding speed $V_{p.sh}$ decreases, the deformation time, and consequently the degree of weakening, increase, and this is bound to lower k and σ_{sr}. The extruding speed, however, has a dual effect: as $V_{p.sh}$ is reduced the metal cools more, and this raises the yield stresses σ_s and σ_{sr}.

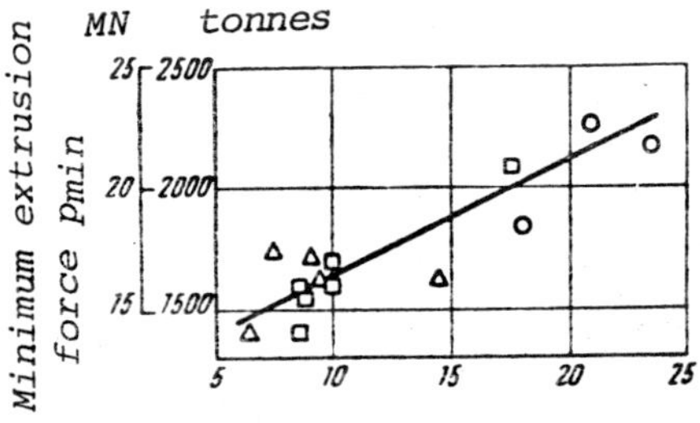

Time spent by billet in container after the moment of back-extrusion τ sec

Fig.41 Relationship of the minimum extrusion force p_{min} to the time spent by the billet in the container after the moment of back-extrusion τ: MNZH5-1 alloy, billet temperature 920-960°C, 3000 tonne press. Container dia. D_k = 306mm, tube dimensions 190x2 4mm, billet lengths L_{sl}: (Δ) 250; (□) 300; (○) 500mm.

Figure 41 shows the relationship of the minimum extruding force P_{min} to the length of time τ spent by the billet in the container after back-extrusion. As τ is increased from 7 to 23 s, when 190 x 204mm MNZh5-1 alloy tubes are extruded the minimum extruding force P_{min} increases from 1500 to 2300 tonnes, i.e. by 53%. Owing to cooling the pressure in the die (p_{min}) may in many cases be higher than the pressure at the beginning of the extrusion process. Figure 42 shows the extruding force with different speeds $v_{p.sh}$ and running billet lengths. When the process has not yet become established, in the initial moment of extrusion (curve 1 in fig.42) increasing the extruding speed by 200-350% raises the extruding pressure by 20-40%. This increase in the pressure is basically due to the increase in the forces of inertia P_{in}. When the process has become established (curves 2), altering the extruding speed within the range indicated increases the extruding force maximum by 5-10%; this has been confirmed by results given in other references (16,44).

When, therefore, the extruding pressure is calculated, the effect of the extruding speed (its variation within the range encountered in practice) can be ignored for the start of the established extrusion process (16). As the process continues to develop the extruding speed affects the cooling of the metal (curves 3 and 4 in fig.42): as the extruding speed decreases the extruding force increases. The higher the strength figures for the metal and the greater the reduction, the more rapidly will the pressure increase by the end of the process; this is due to thermal phenomena.

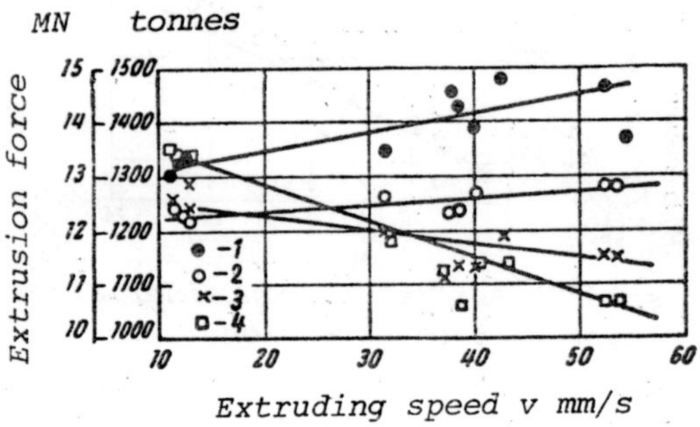

Fig.42 Relationship of the extrusion force P to the main plunger speed v. 1500-tonne press MNZh5-1 alloy billet, dimensions 175x240mm, 55x63mm tube. Running billet length: 1 - 255x284mm, 2 - 216-230mm; 3 - 139-158mm; 4 - 45-70mm.

If we substitute equation (54) in equation (53) and carry out transformations, we get:

$$\frac{p_{min}}{\sigma_s} = A\sqrt{k \ln \mu} \qquad (56)$$

If we take the value of σ_s corresponding to the initial temperature of the billet, we can see from equation (56) that the quantity $A\sqrt{k}$ will be the angular coefficient of the relationship $\frac{p_{min}}{\sigma_s} = \phi(\ln \mu)$ plotted in fig.40. The coefficient k can be determined using ray diagrams of the type plotted in fig.40.

Table 16 contains values of k for different metals and alloys, for $A = 2.5$. Since the relationship $\frac{p_{min}}{\sigma_s} = \phi(\ln \mu)$ is linear, as the parameters of the extrusion process alter within the range encountered in practice the calculation coefficient k is a constant for the alloy under consideration and a specific extrusion temperature.

As the billet length is increased the components T_k and T_{ig} consumed on overcoming the forces of friction along the surfaces of contact between the metal and the container and the mandrel increase. These forces depend on the frictional stresses (τ_k) and (τ_{ig}) along the container and the mandrel, and are equal to:

$$\left.\begin{aligned} T_k &= \pi D_k \, (L_{s1} - h) \, \tau_k \\ T_{ig} &= \pi d_{ig} \, (\psi L_{s1} - h) \, \tau_{ig} \end{aligned}\right\} \qquad (57)$$

Table 16

Values of the coefficient of strengthening k

Metal	Extrusion temperature °C	k	$\sqrt{k}$
Copper	775.950	6.15	2.48
L68 brass	675-825	-	-
L62 brass	675-825	15.4	3.92
LS59 brass	700-750	12.0	3.46
MNZh5-1	900-950	6.7	2.58
MNZhMts30-0.8-1	825-950	5.8	2.4
BrAZhN10-4-4	825-875	2.89	1.7

The specific pressure (stress) on the pressure pad created by the forces of friction ($T_k + T_{ig}$) is equal to:

$$p_t = \frac{T_k + T_{ig}}{\frac{\pi}{4}\,(D_k^2 - d_{ig}^2)} = \frac{4\left(D_k (L_{s1} - h)\,\tau_k + d_{ig}\,(\psi L_{s1} - h)\,\tau_{ig}\right)}{D_k^2 - d_{ig}^2}$$

The coefficient ψ represents the length l_{sk} of the zone in which the metal slides along the mandrel:

$$\psi = \frac{l_{sk}}{L_{s1}}$$

It is pointed out in Chapter II that the coefficient ψ varies between 0.8 and 0.9 for metals and alloys with turbulent flow, and that $\psi = \frac{h}{L_{s1}}$ for laminar flow in the metal, i.e. in the latter case $T_{ig} = 0$.

The frictional stresses along the container were determined from oscillograms as the ratio of the increase in T_k, the force of friction, to the area S of the contact surface of the billet:

$$\tau_k = \frac{\Delta T_k}{\Delta S} = \frac{\Delta T_k}{\frac{\pi}{4} D_k \cdot \Delta L_{s1}} \tag{58}$$

The quantity ΔT_k was defined as the difference

$$\Delta T_k = \Delta P - \Delta T_{ig}$$

where:

ΔT_{ig} = the change in the force of friction along the mandrel accompanying a change of ΔL_{s1} in billet length.

The value of τ_k decreases during extrusion; this is evidently because the increase in the temperature of the tools causes that of the contact layer of metal to rise (fig.43). The frictional stress on the mandrel τ_{ig}, alters in an analogous manner (fig.44).

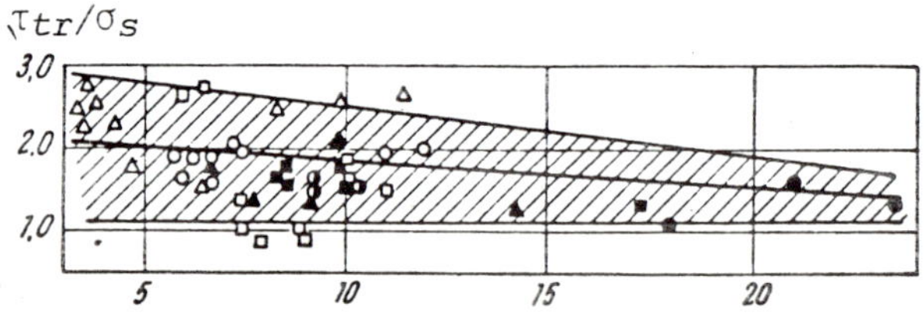

Fig.43 Relationship of the frictional stress in the container to the parameters of the extrusion process: metal MNZh5-1. Billet temperature 900-960°C. 1500-tonne press, Dk=155mm, billet length Lsl: △ - 200; 0 -300; □ -400mm. 3000 tonne press, Dk = 306mm, billet lengths Lsl: ▲ - 250; ■ -300; ◐ 400; ● - 500mm.

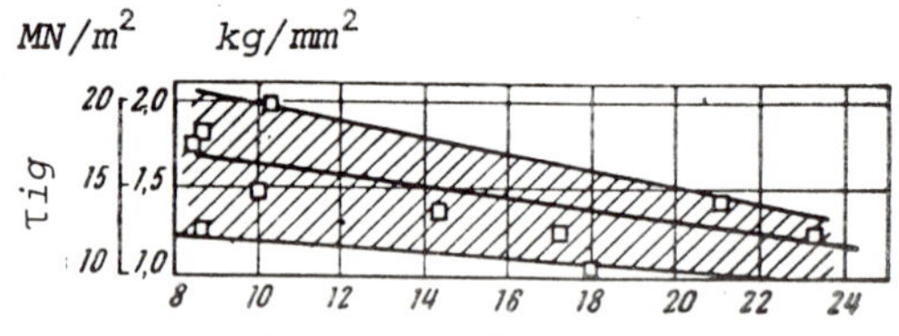

Fig.44 Relationship of the frictional stress in the mandrel to the extrusion parameters. Metal MNZh5-1 billet temperature 900-960°C 3000 tonne press, Dk = 306mm, tube dimensions 190x204mm.

When the time is trebled (from 8 to 24 s) the mean frictional stress in the mandrel decreases during extrusion, in the case of 190 x 204mm tubes, from 17.5 to 12.5MN/m^2 (from 1.75 to 1.25kg/mm^2) i.e. by 1.4%. The fact that the frictional stress decreases more in the mandrel than in the container is due to the lower heat capacity of mandrels. The smaller the diameter of the mandrel the shorter the time taken to heat it to the temperature of the billet, and therefore, as the diameter of the mandrel is reduced the rate at which the stress in the mandrel decreases rises.

It is extremely interesting that the nature (mechanism) of friction along the mandrel (see Chapter II) is the same as that of friction along the container: shear deformations develop in the contact layer. This is confirmed by the relationship of the frictional stress to the deformation time, also by the fact that the frictional stress does not alter when the hydrostatic pressure (the reduction) is increased (see figs.43 and 44).

The force and frictional stress along the container and the mandrel were also determined by extruding billets of different lengths with the other parameters constant (μ, D_k, d_{ig} = *const.*). We can see from fig.45 that, if the length of the billet is doubled, the extruding pressure increases by 25-40%. In this case the value of τ_k was also found from the reduced values of the forces (Table 17), i.e. using relationships similar to that plotted in fig.41.

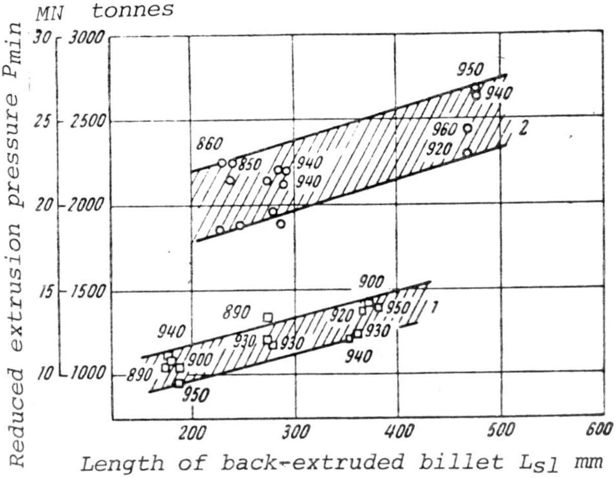

Fig.45

Relationship of the extruding force to the length of the billet. Metal MNZh5-1. Extrusion temperatures 860-950°C (the figures against the points): 1) 1500 tonne press D_k = 155mm tube dimension 40 x 48mm, process time τ = 2.5s; 2) 3000 tonne press D_k = 306mm, tube dimension 190 x 204mm, process time τ = 4.9s.

Table 17: Values of the mean frictional presses τ_k and τ_{ig} along the container and the mandrel, found by extruding billets of different lengths

Press force MN (tonnes)	D_k mm	Tube dimensions $d_{ig} \times D_{tr}$ mm	Processing time τ_s	Extruding forces MN (tonnes)			Friction along mandrel MN (tonnes)		
				P_{max}	P_{min}	$P_{max}-P_{min}$	T_{max}	T_{min}	$T_{max}-T_{min}$
15 (1500)*	155	40 x 48	2.5	13.75 (1375)	10.65 (1065)	3.1 (310)	0.4 (40)	0.12 (12)	0.2 (28
30 (3000)**	306	190x204	4.9	25 (2500)	20.3 (2030)	4.7 (470)	2.4 (240)	1.62 (162)	0.7 (78

* $T_{s1} = 940°C$; $\sigma_s = 21 N/mm^2$ $(2.1 kg/mm^2)$.
** $T_{s1} = 960°C$; $\sigma_s = 19 N/mm^2$ $(1.9 kg/mm^2)$.

Allowance must be made in calculations for the frictional stresses corresponding to the maximum extruding pressure, which in most cases is characteristic of the start of the extrusion process. Since the forces of friction along the container and the mandrel depend on an extremely large number of factors, it can be assumed that:

$$\tau_{ig} = \tau_k = \tau_{tr} = 0.5\,\sigma_s \sqrt{k}$$

This and the assumptions given earlier result in the following expression for the stress in the pressure pad required to overcome the forces of friction along the mandrel and the container:

$$P_t = \frac{4\tau_{tr}\,(L_{s1} - h)}{D_k - d_{ig}} \qquad (59)$$

Friction along container ΔT_k MN (tonnes)	Friction surface area on mandrel ΔS_{ig} mm^2	Friction surface area along container ΔS_k mm^2	L_{max} mm	L_{min} mm	Frictional stresses			
					Along mandrel		Along container	
					τ_{ig} N/mm^2 (kg/mm^2)	$\frac{\tau_{ig}}{\sigma_s}$	τ_k N/mm^2 (kg/mm^2)	$\frac{\tau_k}{\sigma_s}$
2.82 (282)	25 100	93 00	480	280	11.1 (1.11)	0.58	30.3 (3.03)	1.57
3.92 (392)	119 000	192 000	380	180	6.5 (0.65)	0.70	20.4 (2.04)	0.97

The total pressure on the pressure pad can be found using the equations:

$$P = A \sqrt{k}\, \sigma_s \ln \mu + \frac{4\tau_{tr} (L_{sl}-h)}{D_k-d_{ig}} \tag{60}$$

or:

$$P = \sqrt{k}\ \sigma_s \left(A \ln \mu + \frac{2(L_{sl}-h)}{D_k-d_{ig}} \right) \tag{61}$$

It should be noted that, during hot extrusion, the lubricant obviously acts as a heat insulation layer and reduces the degree to which the contact layer of metal cools, consequently also reducing τ_{tr}.

It was pointed out earlier that the value of σ_{sr} also depends on the degree of cooling, which is governed by the thermal conductivity of the metals, the length of time spent by the billet in the container, and the ratio of billet volume contact surface area

Table 18: Tube Extrusion Force

Tube dimensions mm	Container dia. mm	ln μ	Billet length mm	Length h of region of deformation mm	Billet temperature °C	σ_s' N/mm^2 (kg/mm^2)	Extrusion force kN (tonnes^{-1})		Difference %
							Experimental	Found using eq. (64)	
Copper									
15×19	155	5,15	200	82	900	18 (1,8)	1162	1280	+10
15×21	155	4,70	200	82	900	18 (1,8)	1060	1170	+10,4
15×23	155	4,36	200	82	900	18 (1,8)	907	1130	+24
15×27	155	3,85	200	82	900	18 (1,8)	843	944	+12
26×32	155	4,2	200	79	900	18 (1,8)	907	1015	+12
26×34	155	3,88	200	79	900	18 (1,8)	885	951	+7,1
26×38	155	3,4	200	79	900	18 (1,8)	843	856	+1,6
55×59	155	3,83	200	69	900	18 (1,8)	1077	899	—16
55×61	155	3,4	200	69	900	18 (1,8)	1028	821	—20
55×63	155	3,1	200	69	900	18 (1,8)	907	764	—15
55×67	155	2,66	200	69	900	18 (1,8)	901	849	—5,7
20×24	155	5,29	187	80,5	900	18 (1,8)	1050	1222	+16
20×32	155	3,64	187	80,5	900	18 (1,8)	880	887	+0,8
30×34	155	4,52	187	78	900	18 (1,8)	1076	955	—11
34×42	155	3,4	187	77	900	18 (1,8)	8165	835	+2,3
40×52	155	3,06	187	74	900	18 (1,8)	885	761	—14
45×53	180	3,68	240	87	850	20 (2,0)	1470	1371	—6,7
50×58	205	3,9	240	100	750	26 (2,6)	2472	2320	—6,1
50×60	205	2,96	350	100	760	25,2 (2,52)	2330	2108	—9,5
L62 Brass									
30×34	155	4,5	230	77,0	690	16,5 (1,65)	1483	1600	+7,9
35×39	155	4,35	180—190	75,0	730—760	11,5—9,5 (1,15—0,95)	1002—816	1108—860	(+10)—(+5,4)
26×33	180	4,3	190—220	90	750—770	10—9 (1,0—0,9)	1228—988	1200—1082	(—2,3)—(+9,5)
26×34	180	4,15	200—210	90	790—810	8—7 (0,8—0,7)	976—817	897—777	(—8,1)—(—4,9)
45×52	180	3,8	290—300	84	710—790	11,4—7,6 (1,14—0,76)	1332—958	1552—929	(+16,5)—(—3,0)

				LS59-1 Brass					
24×31 20×37 20×50 130×160 130×170	155 180 180 306 306	4,1 3,49 2,74 2,18 1,87	180 180—200 170—180 370 360—410	77,5 89,0 85,0 111,0 109,0	700—710 690—730 620—740 750—820 690—760	8—7,8 (0,8—0,78) 8,2—7,4 (0,82—0,74) 11,2—7,2 (1,12—0,72) 7—5,5 (0,7—0,55) 8,2—6,8 (0,82—0,68)	612—566 730—636 579—495 860—695 917—817	611—597 684—618 781—502 905—713 985—835	(—0,00)—(+5,5) (—6,3)—(—2,8) (+35)—(+0,14) (+5,2)—(+2,6) (+7,4)—(+2,2)
				BrAZhN10-4-4					
90×120 110×140 90×130	306 306 306	2,62 2,39 2,24	350—360 425 300—370	130,0 120,0 125,0	900—910 900 880—900	15—14 (1,5—1,4) 15 (1,5) 17—15 (1,7—1,5)	1400—1250 1323 1439—1232	1480—1370 1550 1500—1230	(+5,4)—(+9,6) +17 (+4,2)—(—0,01)
				MNZh5-1					
35×40 30×38 40×50 65×73 160×174	155 155 180 180 306	4,1 3,7 3,48 3,18 2,63	150—200 190 240—270 250 180—230	74,5 75,0 85,0 75,0 105,0	910—950 970 900—930 920—950 910—920	24—20 (2,4—2,0) 18 (1,8) 25—22 (2,5—2,2) 23—20 (2,3—2,0) 24—23 (2,4—2,3)	1442—1145 1045 1641—1500 1265—1101 2184—1900	134,5—112,0 935 1750—1500 1380—1255 2770—2400	(—6,7)—(—2,2) —10,5 +6,6 (+9,1)—(+14) (+27)—(+26)
				MNZhMts30-0.8-1					
36×52 36×46	155 155	2,79 3,25	200—220 140—150	70,0 72,0	930—990 970—1030	40—32 (4,0—3,2) 35—27 (3,5—2,7)	1334—1160 1553—1370	160,0—125,0 1400—1130	(+20)—(+7,9) (—10,1)— —(—10,2)

(no allowance was made for the contact surface of the die and the pressure pad):

$$\rho = \frac{S_{sl}}{V_{sl}} = \frac{1}{0.25\ D_k\left[1 - \frac{d_{ig}}{D_k}\right]} \tag{62}$$

where:

S_{sl} and V_{sl} are respectively the billet surface contact area along the container and the mandrel and the volume of the billet per unit length of the latter.

We can see from equation (62) that the importance of cooling decreases as D_k is increased and the ratio $\frac{d_{ig}}{D_k}$ reduced. The degree to which the billet metal cools by the moment at which tube extrusion begins can be allowed for by the dimensionless coefficient z, calculated using the empirical equation:

$$z = 1 + \frac{k_T \lambda \Delta T \tau}{D_k\left(1 - \frac{d_{ig}}{D_k}\right)} \tag{63}$$

where:

λ = the thermal conductivity of the metal being deformed at the extrusion temperature;

ΔT = the difference between the temperatures of the heated billet and the tool (container) in °C;

τ = the period between the start of back-extrusion and the beginning of extrusion in s;

k_T = a coefficient equal to $3.04 \cdot 10^{-4}$ with λ in W/m·°C, or $2.6 \cdot 10^{-4}$ with λ in kcal/m·h·°C;

D_k = the diameter of the container in mm.

The pressure on the pressure pad when the process initially becomes stable is:

$$p = \sqrt{k} \cdot z\sigma_s \left(A\ \ln \mu + \frac{2(L_{sl} - h)}{D_k - d_{ig}}\right) \tag{64}$$

The extruding pressure:

$$P = pF = \frac{\pi(D_k^2 - d_{ig}^2)}{4} \cdot p \tag{65}$$

or:

$$P = 0.786 \cdot 10^{-3} \sqrt{k\sigma_S} \cdot z \left[2.5 \; (D_k^2 + d_{ig}^2) \cdot \ln \mu + 2(L_{S1} - h) \; (D_k^2 - d_{ig}^2) \right| \\ (D_k^2 - d_{ig}^2) \Big] \qquad (65a)$$

Values of the extruding force calculated using equation (65) for $z = 1$ are given, together with the true extruding forces, in Table 18.

METHODS OF IMPROVING THE EFFICIENCY OF THE EXTRUSION PROCESS. SELECTION OF RATIONAL BILLET DIMENSIONS

Factors governing press output

The output from equipment is governed by the number of units of production produced per unit of time. If a production unit has a weight g and if n such products are produced an hour in particular plant, production output is governed by the equation:

$$G = \frac{ng}{1000} \text{ tonnes/h}$$

or

$$M = \frac{ng}{m} \text{ m/h;}$$

where m is the weight of a metre run of product.

The expenditure on producing a tonne of products in machine-hours is:

$$C = \frac{1}{ng}$$

and in man-hours it is:

$$Ch = \frac{K}{ng}$$

where K is the number of employees handling the plant.

If the total pay for K employees for an hour is A, the cost of production as to pay is:

$$p = \frac{A}{gn}$$

Naturally the higher the qualifications of the personnel running the plant the higher will be the cost of production.

Shortening the extrusion cycle. The quantity *n* in the above equations is governed by the extrusion cycle, which is equal to the sum of T_{osn} and T_{vs}, the basic and auxiliary times. The basic time consumed on deformation of the metal is governed by the back-extrusion, piercing and extrusion times. In presses with no independent piercing system the basic time is the extrusion time.

We can see from the data given in Tables 19 and 20 that the deformation time comprises only 6-12% of the cycle time. The principal lines followed in shortening the cycle thus consist of the mechanisation and automation of the auxiliary operation.

The piercing and extrusion time is governed by the press speed characteristic. In many cases reduction in the basic time achieved by increasing speeds is limited by the temperature and speed conditions for the process. For instance, we know that when certain alloys (MNZh5-1 type, etc.) are extruded, increasing the extrusion speed causes rapid heating of the metal, and in consequence results in tubes being rejected for overheating, double skin or cracks. In this case the programmed variation of the speed is interesting, also the use of closed automatic regulation systems (tube temperature - extrusion speed) such that the process is carried out under isothermal conditions.

The installation of modern induction heaters and improvements in the system for transporting billets from the heaters to the container and for transporting tubes to the cooler greatly reduce the labour consumed on the operation of heating billets. Modern designs of press (57) have special programme control systems: all the working operations are carried out in a sequence established in advance. When necessary it is possible to go over to push-button control.

The mechanisation of manual operations, also changing to remote push-button control, greatly reduce the number of personnel required to operate plant. Five to seven personnel were required to operate the old design presses, but presses of the new design can be handled by only two or three operators. One exercises control over the press machinery from the control panel, and one or two are engaged on handling the extruded products (57).

The requirement of shortening the extrusion cycle as much as possible has resulted in the development of a design of die holder which is new in principle (57). In the old design process the shears for cutting off the discard were beyond the press (on the receiving tray), and the die holder had to be secured to the moving tray, together with which it moved from the aperture to the front housing, in order to feed the discard beneath the shears. In modern presses, to remove the discard it is not necessary to move

Table 19

Length of the extrusion cycle and billet deformation time for the production of tubes in horizontal and hydraulic presses

Alloys	Billet dimensions $D_{s1} x L_{s1}$ mm	Tube dimensions $d_{tr} x D_{tr}$ mm	Cycle time s	Operation times sec			Deformation time	
				Back-extrusion	Piercing	Extrusion	Seconds	As percentage of cycle time
1500 tonne press								
Copper	150x(200-240)	26x30	51	-	-	-	4.6	9.1
	175x250	26x33	54	-	-	-	4.7	8.7
L62	150x200	30x34	55	-	-	-	6.0	10.9
	175x250	35x44	56	-	-	-	6.4	11.4
MNZh5-1	150x240	35x40	80	-	-	-	7.0	8.7
	175x210	45x55	84	-	-	-	10.0	11.9
LS59-1	175x350	30x60	56	-	-	-	4.7	8.4
3000 tonne press								
Copper	200x250	60x67	103	0.91	1.40	4.04	6.4	6.2
	250x425	90x100	117	0.86	1.50	7.95	10.3	8.8
BAZhN	200x250	40x70	156	1.40	1.70	7.40	10.5	6.7
	300x475	90x130	254	1.90	2.10	10.3	14.3	5.6
MNZh5-1	200x300	40x50	137	0.97	1.35	8.75	11.1	8.1
	250x300	80x90	162	1.00	2.03	7.75	10.8	6.7
	300x300	140x154	176	1.20	2.30	10.20	13.7	7.8

Table 20

Graph of the running of a horizontal 1500 tonne (15MN) hydraulic press used for producing 26x30mm copper tubes from 150x200mm billets (with no control pressure pad)

Operations	Times s
Water cooling of mandrel	9,6
Lubrication of mandrel	1,8
Raising of billet to press axis	4,2
Bringing cotter up to container	3,3
pushing billet into container	2,9
Lowering hoisting equipment	1,2
Raising pressure pad to press axis	2,8
Inserting mandrel into pressure pad aperture	1
Lowering pressure pad hoist	1,5
Lowering wedge seal	4,5
Forcing billet in container up to die	2
Squeezing	0,8
Piercing	1,4
Extrusion	2,8
Raising wedge seal	3,0
Moving die holder	3,0
Reverse travel of cotter	6,0
Positioning slug beneath shears	3,0
Cutting away slug	6,0
Withdrawing die holder from shears	1,3
Positioning extractor	2,7
Extracting product from die	2,4
Raising extractor	0,8
Positioning die holder in container	5,0
Removing slug	4,0
Placing pressure pad on slide	6,0

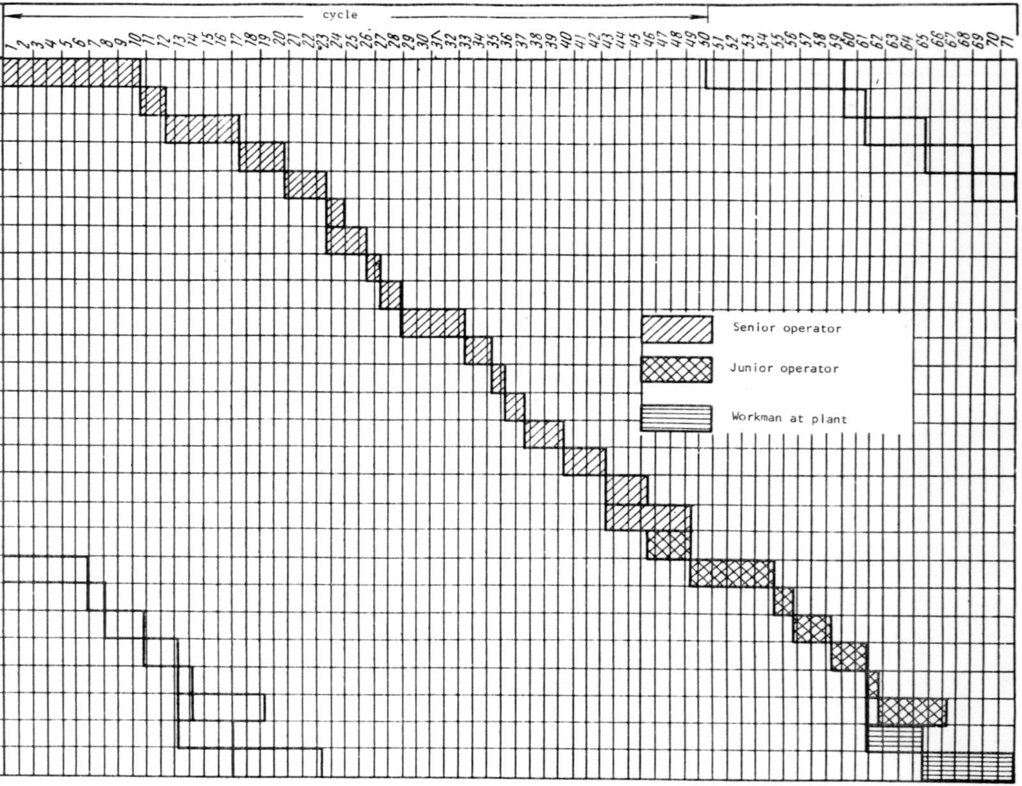

the die holder from the housing aperture, since the shears (or saw) are positioned between the die holder and the container.

Before the operation of removing the discard the container is moved backwards a small distance, and the discard after removal, together with the pressure pad, drops down onto the receiving tray, where the discard is separated from the pressure pad (57). The discards are collected in a box, and the pressure pad is returned down a special chute to the hoisting table at the charging side of the press.

This procedure for removing the discard makes it possible to eliminate the following operations, consequently shortening the cycle time: (1) opening and closing the die holder jaws (there is no mechanism for this in the new presses); (2) moving the die holder and placing it in its initial position. When certain alloys are extruded die life is very short, only amounting to 10-20 extrusion operations; after this the die has to be dressed or changed. The use of new grades of steel for press tools (16) greatly lengthens die life, and this reduces the amount of time consumed on changing dies.

In presses of the old design the dies were usually removed from the die holders by hand, the operation taking between 2 and 5 min. In presses of the new design, in order to reduce this time and make the work of changing tools less laborious, there are special two or four-position heads with two sockets for the dies (57). During extrusion in one position the die in the other position can be cleaned or replaced without interrupting the running of the press. Dies are removed from their sockets in the die holder head by means of a special hydraulic cylinder.

New designs of press, which are the equal of foreign designs as to outward appearance and automation, have been developed in recent years in the Soviet Union.

Increasing billet and tube blank weight. The weight of tube g_{tr} produced in a single working cycle is governed simply by the billet weight g_{sl} and the amount of geometrical waste:

$$g_{tr} = g_{sl} - g_{otkh} \qquad (66)$$

If we divide the left and right hand sides of equation (66) by the specific gravity of the metal of which the tube is made, we obtain:

$$V_{tr} = V_{sl} - V_{otkh} \qquad (67)$$

where V_{tr}, V_{sl} and V_{otkh} are the respective volumes of the tube, the billet and the discard.

Reduction of geometrical losses of metal. The geometrical losses of metal when tubes are extruded are: the discard, the slug formed when a billet is pierced, and the front and rear ends of the tube which are uneven, and in which in many cases the thickness of the tube walls varies greatly. Experimental figures for the size of the slug with relation to the extrusion parameters are plotted in fig.46.

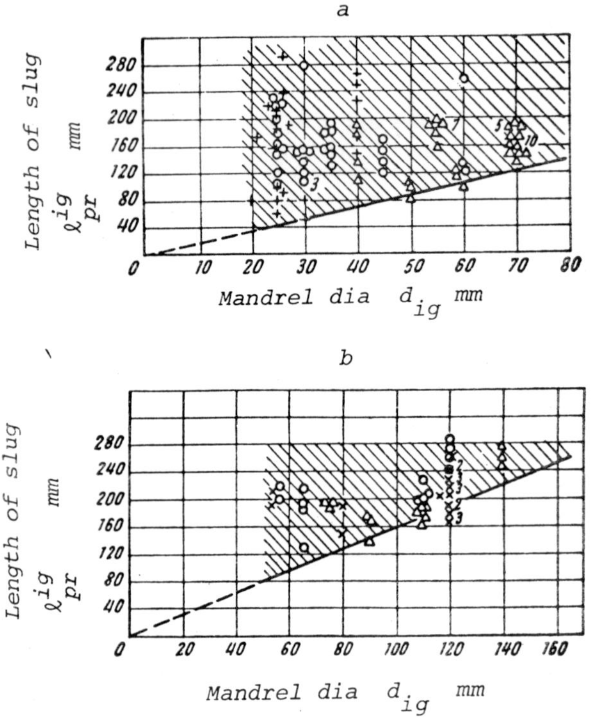

Fig.46: *Relationship of the reduced slug length ℓ_{pr}^{ig} to mandrel diameter d_{ig}. Metal M3S copper: a) container dia. D_k = 180mm billet length L_{sl}: (Δ) - 330-340mm (+) - 250-280mm (0) - 220-230mm b) container dia. D_k = 306mm billet length L_{sl} (Δ) - 370-380mm (0) - 400-420mm (X) - 450mm*

In the absence of a pressure head, i.e. after back-extrusion of the billet, when the pressure is removed from the pressure pad and the latter is withdrawn, a slug of the minimum length is formed (the lower boundary of the shaded region in fig.46). The reduced slug length ℓ_{pr}^{ig} now depends linearwise on the diameter of the mandrel:

$$\ell_{pr}^{ig} = k_1 \cdot d_{ig} \qquad (68)$$

hence the volume of the slug is:

$$V_{prob} = \frac{\pi}{4} d_{ig}^3 \cdot k_1 \qquad (69)$$

Values of the coefficient k_1 are given, for different container diameters in Table 21. These values of k_1 were obtained for $\frac{L_{sl}}{D_k} = 1.0\text{-}1.8$. Obviously as this ratio is reduced and the billet length L_{sl} increased the values of ℓ_{sr}, and consequently also of k_1, are bound to increase owing to rise in the "frontal" resistance. At the same time, with $\frac{d_{ig}}{D_k} < 0.3$, the true length of the slug is much greater than the length calculated using equation (68). The reason for this lies in the practical difficulties for the operator in determining the moment at which the back-extrusion of the billet ends and flow by the latter begins.

Table 21

Values of the coefficient k_1

Press force MN (tonnes)	Container diameter D_k mm	k_1	
		Without pressure head or flow of metal into the die during back-extrusion of the billet	In most cases
15 (1500)	155	-	2-4 or more
	180	1.8	1.85
30 (3000)	205	1.75	1.8-4.0 or more
	255	1.70	1.7-2.6
	306	1.60	1.6-2.0
	408	1.50	1.5-1.7

When large diameter tubes are extruded the volume of waste metal due to piercing increases abruptly (Table 22). To reduce the "slug" the measures employed are either preliminary piercing in special auxiliary piercing presses (5,23) or basically piercing the billet in the container using a "blind" (no aperture) die (this is called the combined method). In the latter case, if the press does not have a multi-position die holder the operation of extrusion is carried out in the following sequence:

Table 22

Distribution of waste metal percentages by types

$\ell_{pr}^{ig} = 1.6\ d_{ig},\ \ell_{ost} = 0.12\ D_k,\ t_{rub} = 2mm,\ \ell_{obr} = 0.5m$

Tube dimensions mm	Billet		End scalp		Slug		Scalp		Discard		Total amount of lost metal	
	Dimensions mm	Volume cm^3	cm^3	%*	cm^3	%*	cm^3	%*	cm^3	%*	cm^3	%**
Copper												
55 x 100	300 x 450	31 800	208	3.8	2610	47.6	-	-	2730	49.6	5 548	17.4
160 x 175	300 x 450	31 800	5130	56.6	1965	21.7	-	-	1970	21.7	9 065	28.5
110 x 123	300 x 380	26 800	1670	32	2350	45.2	-	-	1190	22.8	5 210	19.4
190 x 204	300 x 400	28 200	8600	69.2	1660	13.4	-	-	2170	17.4	12 430	44.2
MNZh5-1												
30 x 61.5	150 x 200	3530	38.1	2.0	338	20.0	192	11.3	1130	66.7	1694	48.0
40 x 66.5	150 x 200	3530	80.3	4.7	326	19.1	192	11.3	1110	65.0	1708	48.4
60 x 78.5	150 x 200	3530	27.1	15.3	298	16.9	192	11.3	1010	51.0	1771	50.2
40 x 48	150 x 200	3530	80.3	9.2	326	37.2	192	11.3	276	31.6	874	24.8

* Percentage of total amount of waste metal

** Percentage of initial volume of billet

(1) the container is closed by the die holder with the die corresponding to the tube to be extruded;

(2) the wedge seal is closed and the plug (the "blind" die) is introduced into the container by means of a cotter. When cold, the diameter of the plug is 2-3mm less than that of the container liner. Before the start of extrusion the plug and die are heated to 300-350°C to reduce the extent to which the billet is cooled during piercing;

(3) the billet and the tube pressure pad, with the required dimensions, are introduced into the container;

(4) the billet is pierced, and the cotter is withdrawn. At the end of piercing a 10-15mm thick layer of metal remains between the end of the mandrel and the plug. The billet is therefore pierced in a manner similar to the reverse extrusion method;

(5) after piercing the wedge seal is releaed, and the head, together with the die holder and die are removed from the press throat;

(6) pressure is fed into the mandrel cylinder and the remaining layer of metal is pierced (cut). The plug is forced out of the container by the mandrel. Unpierced metal is withdrawn from the pressure pad by means of a steel hook;

(7) the head, together with the die holder and die, are re-introduced into the press throat. When the wedge seal has been closed the extrusion process is carried out in the normal manner.

Table 23 gives the yields of sound metal when copper tubes are extruded by the normal and combined methods. We can see from the figures in this table that piercing with a plug increases the yield of sound metal from 60.4 to 83.2% when tubes with dimensions of 260 x 280mm are extruded, and reduces the length of the billet by 150mm.

The extent of the slug can be found using the equation:

$$V_{ost} = \ell_{ost} \frac{\pi}{4} (D_k^2 - d_{ig}^2) \quad (70)$$

where:

D_k and d_{ig} = the diameters of the container and the mandrel;

ℓ_{ost} = the length of the slug.

The size of the slug is governed by the following factors:

(1) the maximum press force. The increasing importance of the forces of friction acting along the surface between the pressure

pad and the metal and between the die (die holder) and the metal, also the cooling of the metal in the "dead" zones against the die and the pressure pad cause considerable increase in the extruding force with $\frac{\ell_{ost}}{D_k} < 0.05$;

Table 23

Yield of sound metal when 260 x 280 x 5000mm copper tubes are extruded in a 3000 tonne hydraulic press with a container with D_k = 408mm

Method of piercing	Billet		End scalp		Slug		Discard		Yield of sound metal %
	Size mm	Vol. cm^3	Length mm	Vol. cm^3	Length mm	Vol. cm^3	Length mm	Vol. cm^3	
In die	400x550	69000	390	20700	50	3880	400	2770	60.4
With plug	400x400	50200	15	796	50	3880	400	2770	83.2

(2) the formation of a laminated region. When certain metals and alloys (for instance the MNZhMts30-0.8-1) are extruded, a laminated region forms at the end of the deformation process: surface layers of the billet enter the metal. Owing to contamination with lubricant and scale, also to cooling of the surface layers, the metal does not weld together and lamination develops. The length of plug corresponding to the moment at which the laminated region forms depends on the type of lubricant used, the temperature drop between the tools and the billet, also the degree of deformation (reduction) and the physical properties of the metal being extruded.

Table 24 gives calculated (from technological charts) and true values of the slugs with relation to the dimensions of the tubes, the metal of which they are made, and the diameter of the container. The length of the slug can be found by the equation:

$$\ell_{ost} = k_2 D_k \quad (71)$$

where k_2 is a coefficient values of which are given in Table 24. When extrusion is carried out using different containers, it can approximately be assumed, for most metals and alloys, that $k_2 \approx 0.12$

The length of tube cut from the forward and rear ends of the tube, which is discarded, is governed by the fluctuation in tube

Table 24 Extent of the slugs (in mm) according to the technological instructions when tubes are produced in horizontal hydraulic presses

Press	Container diameter mm	Metals: Copper	Metals: MnZh5-1 LN65.5	Metals: Aluminium and its alloys	Metals: LS59-1 LO62-1 LZhMts 59-11 L62	Metals: BrA5 L68 LO70-1	Metals: BrAZhN-10-4-1
1500-m	155	30—35 (15—25) / 0,19—0,23 (0,1—0,16)	25—30 (25—30) / 0,16—0,2	20—25 / 0,13—0,17	30—35 (20—25) / 0,19—0,22 (0,13—0,16)	—	—
	180	**30—35 (15—25) / 0,17—0,19 (0,08—0,04)**	30—35 (30—35) / 0,17—0,19	25—30 / 0,14—0,17	35—40 (25—35) / 0,19—0,22 (0,14—0,19)	—	—
3000-m	205	30—35 / 0,15—0,17	30—35 / 0,15—0,17	25—30 / 0,12—0,17	35—40 / 0,17—0,20	30—35 / 0,15—0,17	50—55 / 0,25—0,27
	255	30—35 (25—35) / 0,12—0,14 (0,1—0,14)	40—45 / 0,16—0,18	30—35 / 0,12—0,14	35—40 / 0,14—0,16	—	50—55 / 0,20—0,21
	306	30—35 (30—35) / 0,1—0,11	40—45 / 0,13—0,15	35—40 / 0,11—0,13	45—50 (30—35) / 0,15—0,16 (0,1—0,11)	—	50—70 / 0,16—0,23
	408	35—40 (30—40) / 0,09—0,1 (0,08—0,1)	40—45 / 0,1—0,11	35—40 / 0,09—0,1	50—60 (45—50) / 0,12—0,15 (0,11—0,12)	—	70—75 / 0,17—0,18

Note *The figures in parentheses are mean statistical figures for the true values of slug sizes (the upper figures) and coefficient*

wall thickness at the start of extrusion, when the process is unsteady, and the possible laminated region at the end of the process. When tubes are extruded considerable amounts of metal are wasted owing to fluctuations in thickness, and this reduces the yield of sound tubes. Most authors (3,5,20 etc) agree that the principal reason for fluctuations in tube wall thickness are incorrect assembly of the press and incorrect positioning of the tool.

The fluctuation in tube wall thickness is determined using the equation:

$$\phi = \frac{\Delta t}{t_{tr}} \cdot 100\% = \frac{t_{tr} - t_{min}}{t_{tr}} \cdot 100\% \qquad (72)$$

where t_{tr} is the tube wall "completeness", and is equal to:

$$t_{tr} = \frac{t_{max} + t_{min}}{2} \qquad (73)$$

The wall thickness fluctuation at the leading end always decreases abruptly along the tube towards its centre. The reason is that, in the initial moment of extrusion, the conditions are highly unfavourable to self-centering of the mandrel, and during piercing this may wander to one side. Under normal press working conditions, the extent of this depends on many factors, particularly the extent of back-extrusion of the billet, the effectiveness of heating, local defects in the metal, the quality of the billet ends, etc.

At the moment when extrusion begins the mandrel tends to occupy a steady position relative to the extrusion axis, and this corresponds to equilibrium conditions. The flow of metal now moves the mandrel in the direction of the greatest gap between it and the die aperture. The self-centering of the mandrel depends on the diameter and length of the mandrel region in the container, i.e. from the base of the pressure pad to the die aperture. The larger the diameter of the mandrel and the smaller its length, the greater will be the force required for it to self-centre. As the diameter of the mandrel is increased, therefore, the fluctuation in wall thickness at the leading end of the tube increases.

At the end of the extrusion period, owing to reduction in the length of mandrel in the container the conditions for its self-centering deteriorate and its axis occupies a position close to the true centering of the tool. The fluctuation in thickness at the rear end therefore decreases much more slowly towards the centre of the tube than the fluctuation in thickness at the leading end.

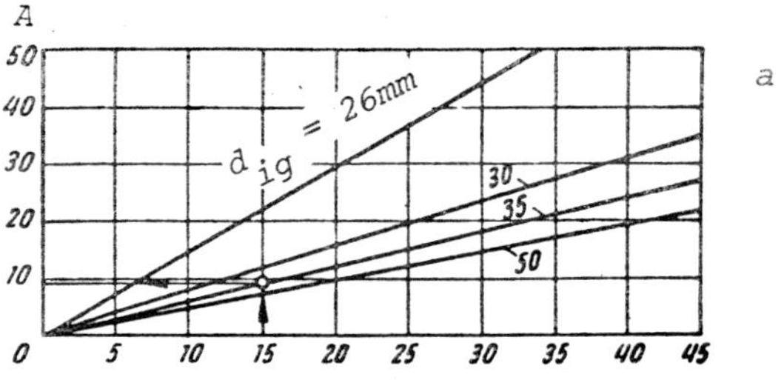

Tube wall thickness fluctuation φ %

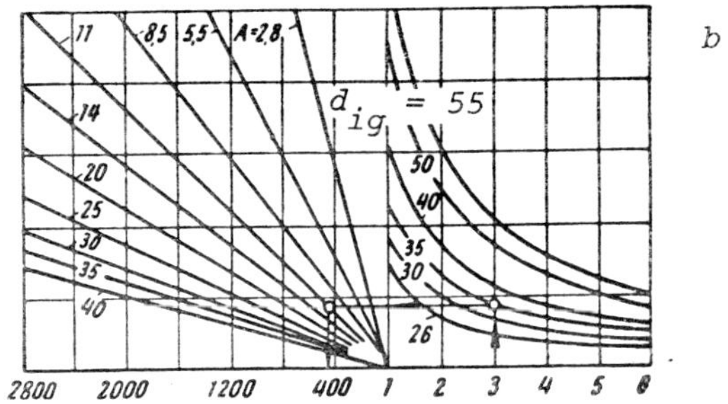

Length of discard from leading end of tube ℓmm *Tube wall thickness t_{tr} mm*

Fig. 47: Nomograms for determining: a) the coefficient A
b) the length of tube from the leading end, in which the fluctuation in wall thickness exceeds the permissible fluctuation

Self-centering of the mandrel in the equilibrium position depends basically on its diameter d_{ig}, the tube wall thickness t_{tr}, the container diameter D_k, the billet length L_{sl}, and the length of tube from the leading end ℓ. Variations in the diameter of the container and the length of the billet do not have any great effect on variations in tube wall thickness from the leading end. Reference (88) established the following relationship of fluctuations in tube wall thickness from the leading end:

$$\phi = A \cdot \frac{d_{ig}^2}{t_{tr} \cdot \ell} \qquad (74)$$

here A is a coefficient representing the variation in the initial fluctuations in tube wall thickness over ℓ = 300mm. The value of the coefficient and the length of tube from the leading end with increased fluctuations in thickness can be determined from the nomograms given in reference (88) (fig.47). The sequence in which the length of discard from the leading end is determined is as follows:

(1) the permissible fluctuation ϕ in tube wall thickness is specified and the value of the coefficient A is determined from the nomogram (fig.47a). For instance, in the case of extrusion with a mandrel with d_{ig} = 35mm and ϕ = 15%, we find that $A \sim 10$;

(2) we use the nomogram in fig.47b for determining the length of tube (discard) in which the fluctuation in wall thickness exceeds 15%. In the case we are considering, for t_{tr} = 3mm we find ℓ_{obr} = 450mm. The volume of the discard can be found using the equation:

$$V_{obr} = \frac{\pi}{4} (D^2_{tr} - d^2_{tr}) \ell_{obr} \quad (75)$$

where ℓ_{obr} is the length of tube cut from the leading and rear ends of the tube.

When a specific range of tubes are produced (brass and bronze), extrusion is carried out with a "scalp", which is also discarded. The billet surface coated with scale and lubricant does not form part of the tube surface (5). The volume of the "scalp" is:

$$V_{rub} = \frac{\pi}{4} (D^2_k - D^2_{sh}) (L_{sl} - \ell_{ost}) = \pi(D_k - t_{rub}) \cdot t_{rub} (L_{sl} - \ell_{ost}) \quad (76)$$

here:

D_{sh} = the diameter of the pressure pad;
t_{rub} = the thickness of the scalp wall;
L_{sl} = the length of the back-extruded and pierced billet:

$$L_{sl} \approx L_{zag} \frac{D^2_{zag}}{D^2_k}$$

The thickness t_{rub} of the scalp is usually equal to 1-2mm. Thus the total volume of the geometrical waste metal is:

$$V_{otkh} = V_{prob} + V_{ost} + V_{obr} + V_{rub} = k_1 \cdot \frac{\pi}{4} d^3_{ig} + k_2 \cdot \frac{\pi D_k}{4} (D^2_k - d^2_{ig}) + \frac{\pi}{4} (D^2_{tr} - d^2_{ig}) L_{obr} + \pi(D_k - t_{rub}) t_{rub} (L_{sl} - \ell_{ost}) \quad (77)$$

Table 22 gave the percentages of each of the items in equation (77). The yield G of sound metal during extrusion is:

$$G = \frac{V_{tr}}{V_{sl}} = 1 - \frac{V_{otkh}}{V_{sl}} \quad (78)$$

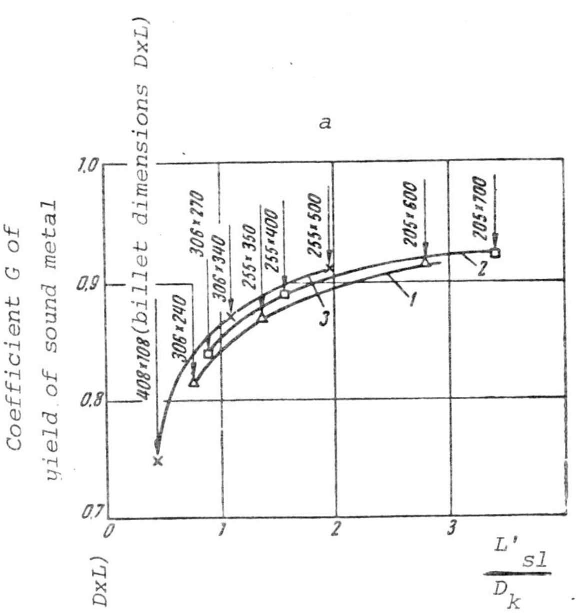

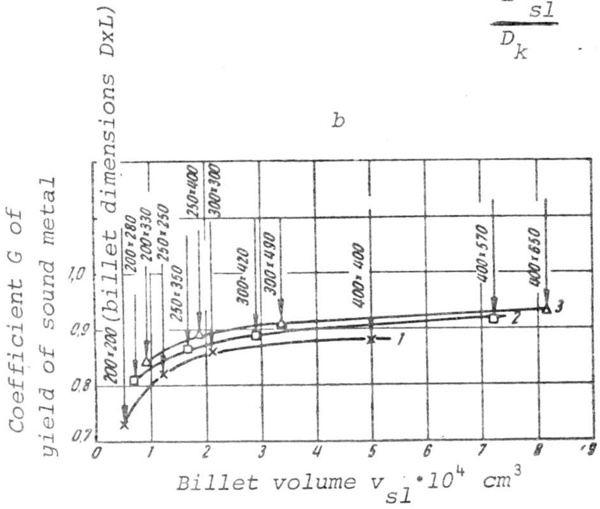

Fig.48: Relationship of G the yield of sound metal when 70 x 84mm copper tubes are extruded in a 3000-tonne press:

a) to the ratio $\frac{L'_{sl}}{D_k}$ with V_{sl} = const.

1) billet volume 16 500cm^3; 2) 18 800cm^3 3) 23 500cm^3

b) to billet volume v_{sl} with $\frac{L'_{sl}}{D_k}$ const.

1) $\frac{L'_{sl}}{D_k}$ = 1; 2 -1.4; 3 -1.6

The reduction in the ratio $\frac{V_{otkh}}{V_{sl}}$ and increase in billet mass (V_{sl}) result in increase in tube mass, and consequently in improvement in the economic effectiveness of the process. We can see from the diagram in fig.48a that, with V_{sl} = const., as the ratio $\frac{L_{sl}}{D_k}$ increases the coefficient of yield of sound metal (G) increases. Increase in billet volume with $\frac{L_{sl}}{D_k}$ = const. also leads to this (fig.48b).

The increase in billet volume is limited by the following factors:

(1) the maximum force developed by the press;
(2) the length of the press receiving table (cooler);
(3) the strength of the mandrel during piercing of the billet;
(4) the physical properties of the metal and the extrusion process conditions, which may on the one hand limit the maximum extruding speed, and may on the other hand cause the billet to cool in the container through increasing the deformation time.

In cases in which the extruding forces are not limited by the power of the press and the physical properties of the metal permit a wide range of variation in the extruding speed, the maximum tube length is governed by that of the receiving table and press cooler L_{st}. If an extruded tube with a length L_{tr} is cut into n parts with lengths ℓ_{tr},

$$L_{st} \geqslant L_{tr} + \ell_{obr} = n\ell_{tr} + \ell_{obr} \tag{79}$$

With ℓ_{obr} = const., the greater the length of the extruded tube the smaller will be the discard losses:

$$G' = 1 - \frac{\ell_{obr}}{L_{tr}} \tag{80}$$

With a limiting table length, the greater the length of the billet, the higher the yields of sound metal G (78) and G' (80) (Fig.48a, V_{sl} = const.). However the minimum extruding forces only correspond fully to the specific ratio $\frac{L_{sl}}{D_k}$ with V_{sl} = const.

The minimum force for extruding a tube of a prescribed dimension from a billet of a particular volume (V_{sl} = const.) can be found by the successive substitution of different values of L_{sl} in equation (65). To do this, the following relationship must be used as the basis:

$$F_{sl} = \frac{V_{sl}}{L_{sl}}; \quad \mu = \frac{V_{sl}}{L_{sl} \cdot F_{tr}} \quad \text{and} \quad D_k = \sqrt{\frac{4V_{sl}}{\pi L_{sl}} + d_{ig}^2}$$

It follows, however, from what has been stated above, that it is economically advisable to extrude from a longer billet if the force developed by the press and its speed characteristic permit this. In practice, therefore, the question of selecting billet length is usually solved by successive approximation to $k[P_{pr}]$, where k is the press force margin factor, equal to 0.85-0.90, and $[P_{pr}]$ is the nominal maximum force developed by the press.

Billet length is governed by the longitudinal bending strength of the mandrel. The smaller the diameter of the mandrel the higher the stresses in it during piercing (16.26), and consequently the smaller the possible billet length. A procedure for calculating maximum billet length on the basis of the longitudinal bend strength of the mandrel was given in refereneces (16,34).

A procedure for working out a technology (schedule) for extruding tubes of particular dimensions, providing the most efficient process possible, is given below.

Procedure for calculating billet dimensions and technological parameters

We will use an example for considering the sequence of calculation. The initial data are: tube metal - copper; billet temperature T_{sl} = 900°C; tube dimensions 100 x 128 x 3000mm.

1. We select the container diameter;

(a) we approximately specify a container diameter D_k = 306mm, and determine the reduction:

$$\mu = \frac{F_{sl}}{F_{tr}} = \frac{(D_k - d_{ig})\ (D_k + d_{ig})}{(D_{tr} - d_{ig})\ (D_{tr} + d_{ig})} =$$

$$\frac{(306-100)\ (306+100)}{(128-100)\ (128+100)} = \frac{206 \cdot 406}{28 \cdot 228} = 13.1$$

hence:

$$\ell n\ \mu = 2.3 \log 13.1 = 2.3 \cdot 1.17 = 2.69;$$

(b) we determine the specific pressure p_{min} on the pressure pad using equation (56);

$$p_{min} = 2.5\ \sigma_S \quad \sqrt{k}\ \ell n\ \mu$$

here, from the handbook σ_S = 1.8kg/mm^2 (18N/mm^2), and from Table 16, $\sqrt{k} = 2.48$.

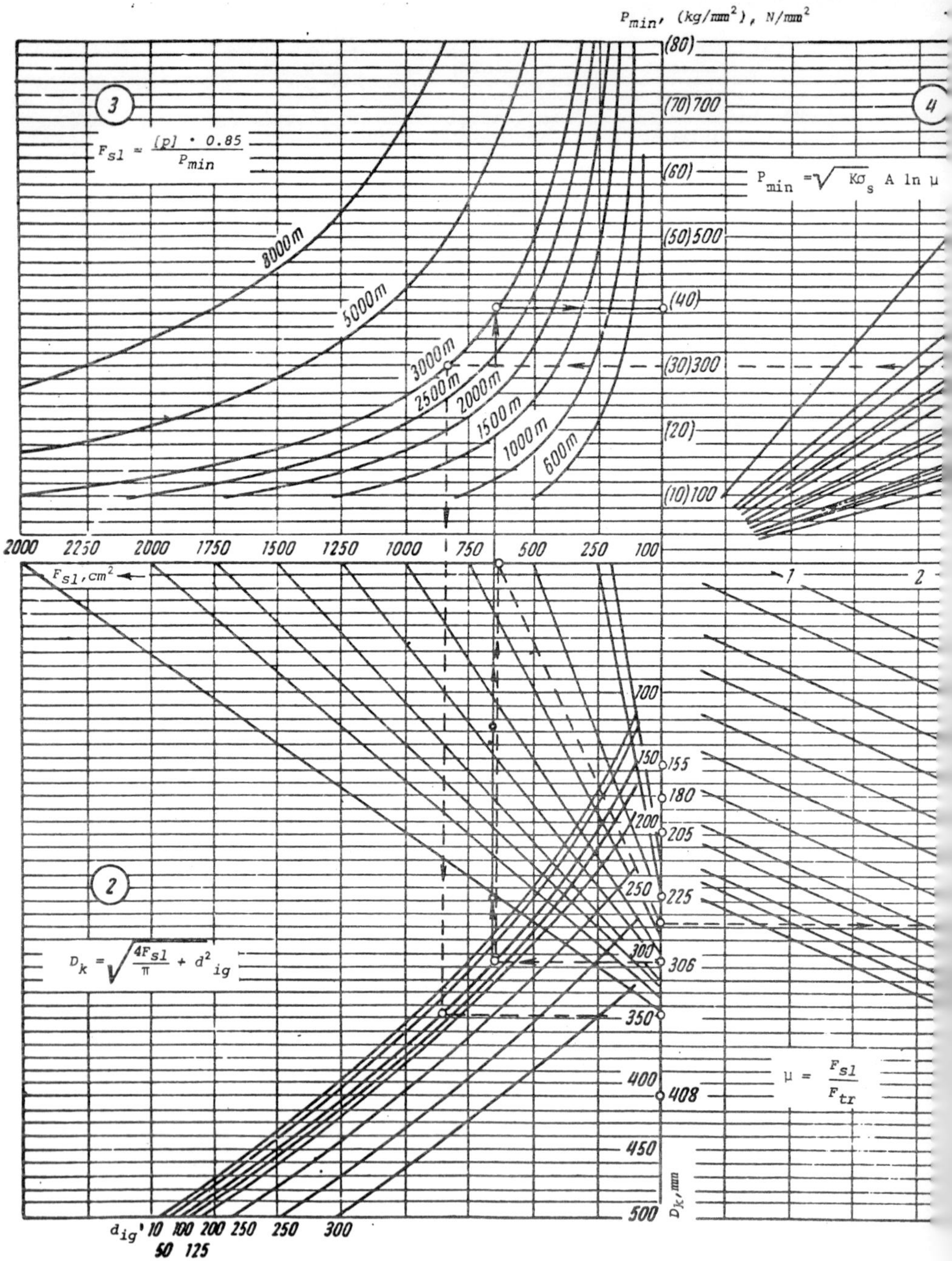

Fig.49 Nomogram for determining the diameter of the container for t

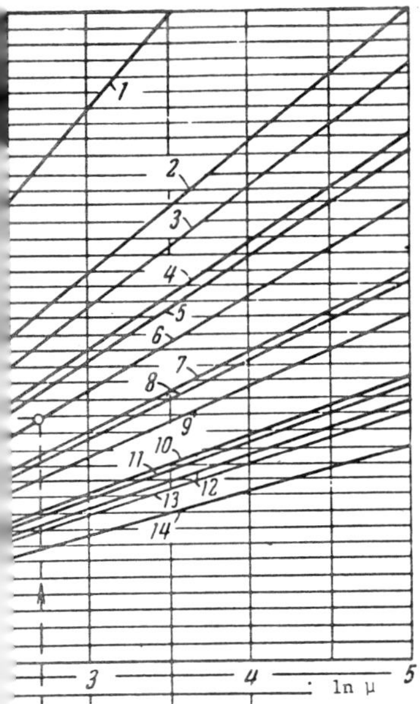

Metal	Temperature, °C					
	700	750	800	850	900	950
MNZh5-1				1	2	4
L62	3	7	11			
Copper				5	6	8
LS59-1	10	13	14			
BAZhN10-4-4				9	12	

Example: Specified: d_{ig} = 100mm, t_{tr} = 14mm
(1) From "a" we find: $F_{tr} \approx 50cm^2$
(2) We specify a container diameter D'_k = 306mm. 3000 tonne press. Successively using nomograms 2, 5, 4 and 3, then again 2 (the interrupted line) we find D''_k = 345mm.
We find the initial diameter D_k from the condition:
$D'_k < D_k < D''_k$
(3) We take the dimension $D_k \approx$ 306mm as the next lower of the standard series of containers.
(4) For the type of container found above, the specific pressure developed by the selected press is determined from nomograms 2 and 3 (the continuous line): P_{min} = 39.5kg/mm² or 395N/mm²
(5) The specific extruding pressure is determined using nomograms 5 and 4 (the interrupted line) P_{min} = 30.0kg/mm² or 300N/mm²

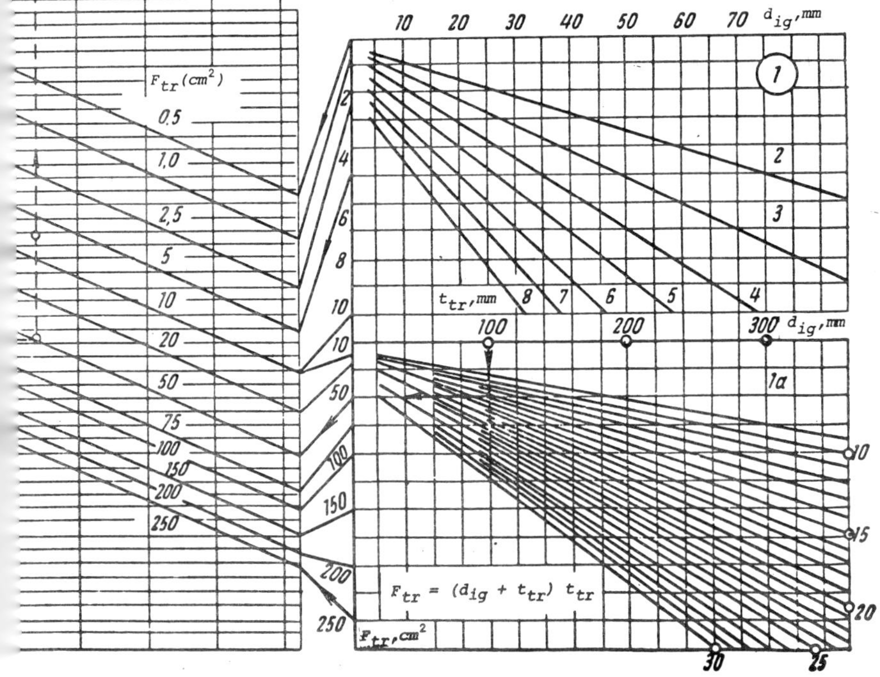

extrusion.

$$p_{min} = 2.5 \cdot 1.8 \cdot 2.48 \cdot 2.69 = 30.1kg/mm^2 \ (301N/mm^2)$$

(c) we determine the billet area on the basis of the maximum force exerted by a 3000-tonne press with a force margin factor of 0.85:

$$F_{sl} = \frac{0.85[p_{pr}]}{p_{min}} = \frac{0.85 \cdot 3\ 000\ 000}{30.1} = 84\ 800mm^2$$

Hence the diameter of the container for p_{min} is:

$$D_k = \sqrt{\frac{4F_{sl}}{\pi} + d_{ig}} = \sqrt{\frac{4 \cdot 84\ 800}{3.14} + 10\ 000} = 345mm$$

We select the next smaller container from the normal range: D_k = 306mm. This calculation can be greatly simplified by using the nomograms in fig.49.

We use nomogram 1 (or 1a) to determine the tube area: with d_{ig} = 100mm and t_{tr} = 14mm, the area F_{tr} = 50cm^2. On the D_k scale we then set a container diameter D'_k = 306mm, and successively use nomograms 2, 5 and 3, then again 2. We do not record the intermediate values, and obtain a value D''_k = 345mm on the D_k scale. Consequently the container diameter D_k which we require must satisfy the condition:

$$D'_k \leqslant D_k < D''_k$$

In the case of considerable discrepancy between D''_k and D'_k (if several container dimensions in the normal range lie between them), the calculation is repeated with the condition:

$$D'''_k \approx \frac{D'_k + D''_k}{2}$$

2. We now calculate the specific pressure developed by the press with D_k = 306mm.

$$[p] = \frac{0.85[p_{pr}]}{F_{sl}} = \frac{0.85 \cdot 3\ 000\ 000}{(306-100)\ (306+100)} = 38.8kg/mm^2 \ (388N/mm^2)$$

On the basis of D_k = 306 (D_k scale), using the nomogram in fig. 49 and successively employing Parts 2 and 3 on the p_{min} scale, we find:

$$[p] = 39kg/mm^2 \ (390N/mm^2)$$

3. We determine the pressure margin Δp

$$\Delta p = [p] - p_{min} = 38.8-30.1 = 8.7kg/mm^2 \ (87N/mm^2)$$

p_{min} is found from the nomogram in fig.49 in the following manner:

We specify a value of 306mm on the D_k scale, and use Parts 2, 5 and 4 successively on the p_{min} scale; this provides a value of 30.3kg/mm^2 (303N/mm^2). Hence:

$$\Delta p \approx 39.0-30.3 = 8.7kg/mm^2 \quad (87N/mm^2)$$

The pressure margin Δp must satisfy the condition:

$$\Delta p = (0.25-0.3)\ p_{min} \tag{81}$$

With a pressure margin of zero (Δp = 0), extrusion can be carried out using a billet with a length equal to that of the deformation region zone. If, therefore, condition (81) is not fulfilled the next lower container diameter must be employed.

4. We determine the billet length, assuming that Δp is consumed on overcoming the forces of friction along the billet-mandrel and billet-container contact surfaces i.e. $\Delta p = p_t$ and

$$L_{sl} = \frac{\Delta p\ (D_k - d_{ig})}{2\sqrt{K}\ \sigma_s} + h \tag{82}$$

where h is the length of the region of deformation, determined using equation (52), or to simplify the latter we get:

$$h = \frac{D_k}{2\ \sin\ \alpha} - \frac{D_{tr}}{2\ \tan\ \alpha} = \frac{306}{2\ \sin\ 60} - \frac{128}{2\ \tan\ 60} =$$

$$\frac{306}{2\cdot 0.865} - \frac{128}{2\cdot 1.73} = 140mm$$

Hence:

$$L_{sl} = \frac{8.7\ (306-100)}{2\cdot 2.48\cdot 1.8} + 140 = 346mm$$

We use the nomogram in fig.50 to find $\frac{d_{ig}}{D_k} = \frac{100}{306} = 0.327$, and accordingly h = 140mm. From the nomogram in fig.51 we find $L_{sl} - h$ = 206mm, whence L_{sl} = 206 + 140 = 346mm.

5. We determine the geometrical volume of scrap metal extrusion:

(a) from Table 21 we find, for d_{ig} = 100mm(D_k = 306mm), k_1 = 1.6; hence the volume of the end scalp

$$V_{prob} = \frac{\pi}{4} k_1 d_{ig}^3 = \frac{\pi}{4} \cdot 1.6 \cdot 10^3 = 1256cm^3;$$

(b) with k_2 = 0.12, the volume of the slug is:

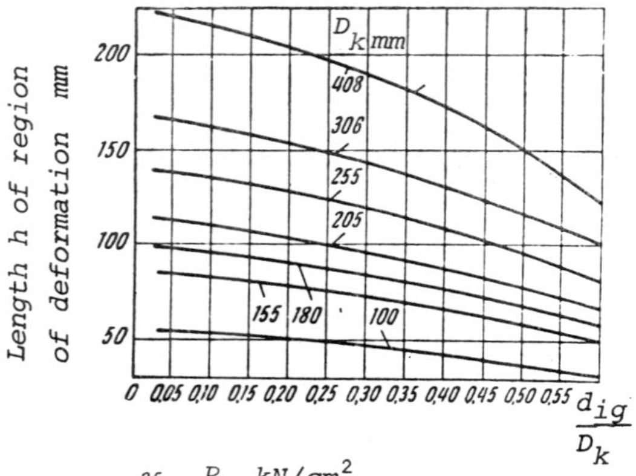

Fig.50: Nomogram for determining the length h of the region of deformation.

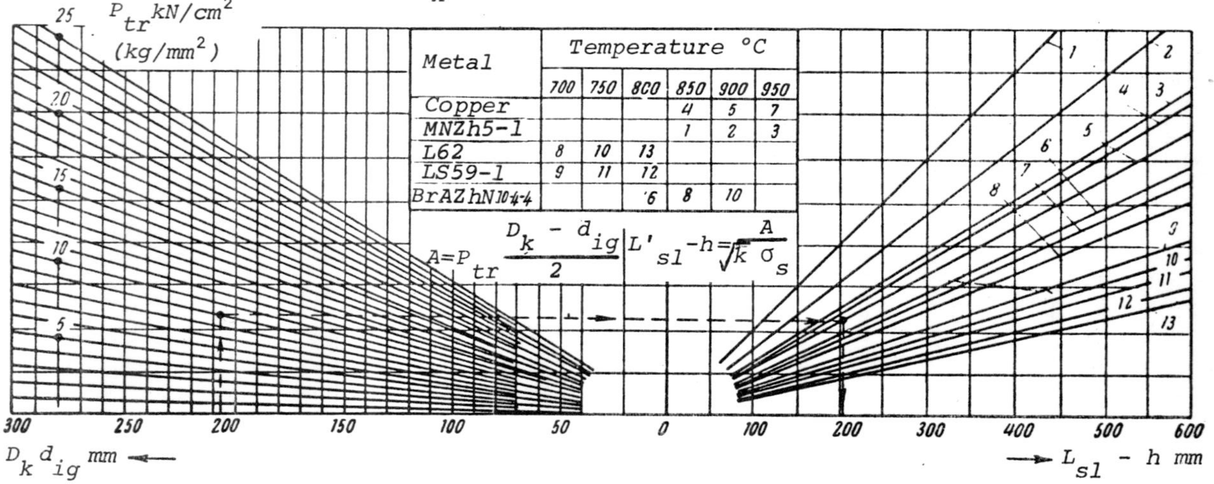

Fig.51: Nomogram for determining billet length L_{sl} in tube extrusion

$$V_{ost} = \frac{\pi}{4} D_k (D_k - d_{ig}) (D_k + d_{ig}) K2 = 2420cm^3.$$

(c) the volume of the discard with $\ell_{obr} = 500mm$, is

$$V_{obr} = \frac{\pi}{4} (D_{tr} - d_{ig}) (D_{tr} + d_{ig}) \ell_{obr} = 2510cm^3;$$

(d) the total volume of discarded metal is:

$$V_{otkh} = 1256 + 2420 + 2510 = 6186cm^3.$$

6. We find the volume of the billet and the tube blank:

$$V_{sl} \doteq \frac{\pi}{4} (D_k - d_{ig}) (D_k + d_{ig}) L_{sl} = 658.46 \cdot 34.8 = 22\ 900cm^3,$$

$$V_{zag} = V_{sl} - V_{otkh} = 22\ 900 - 6186 = 16\ 714cm^3.$$

7. We determine the number of tubes which can be produced from a particular billet:

(a) tube volume:

$$V_{tr} = \frac{\pi}{4} (D_{tr} - d_{ig}) (D_{tr} + d_{ig}) L_{tr} = 15\ 048cm^3;$$

(b) the number of blanks per billet:

$$n_r = \frac{V_{sl}}{V_{tr}} = \frac{16\ 714}{15\ 048} = 1.11$$

Thus one tube can be extruded from the selected billet, i.e. $n_f = 1.0$ (rounding off n_r to the nearest lower whole number). In the case of $\frac{n_f}{n_r} \geqslant 0.8$, n_r can be rounded off to the nearest higher round number.

8. We find the pierced billet length L_{sl} more accurately for $n_f = 1.0$:

$$V_{sl} = n_f V_{tr} + V_{obr} + V_{ost} = 15\ 048 + 2\ 510 + 2420 = 19\ 978cm^3$$

hence:

$$L_{sl} = \frac{4V_{sl}}{\pi (D_k - d_{ig})(D_k + d_{ig})} = \frac{4 \cdot 19\ 978}{\pi \cdot 20.6 \cdot 40.6} = 30.3cm = 303mm$$

9. We determine the volume of the blank and the length of billet back-extruded but not pierced:

$$V_{zag} = n_f V_{tr} + V_{obr} + V_{ost} + V_{prob} = 15\ 048 + 2510 +$$

$$2410 + 1256 = 21\ 234cm^3$$

$$L_{sl} = \frac{4V_{zag}}{\pi D_k^2} = \frac{4 \cdot 21\ 234}{\pi \cdot 30.6^2} = 28.8cm = 288cm$$

The length of the initial billet-blank is:

$$L_{zag} = L_{sl} \frac{D_k^2}{D_{zag}^2} = 288 \cdot \frac{306^2}{300^2} = 300mm$$

The dimensions of the billet are therefore 300 x 300mm.

If the billets arrive from the foundry with a length tolerance of $\pm a$ mm, the length of the billet-blank must be taken as $L_{zag} + a$ and this length must be used as the basis for further calculations.

10. We determine the maximum extrusion pressure:

(a) the pressure at the start of extrusion is determined using equation (64) for the value of $\sqrt{k}$ found from Table 16:

$$p = \sqrt{k}\ \sigma_s \left[2.5\ \ln\mu + \frac{2(L_{sl} - h)}{D_k - d_{ig}}\right] = 2.48 \cdot 1.8$$

$$\left[2.5 \cdot 2.67 + \frac{2(303 - 140)}{306 - 100}\right] = 36.8kg/mm^2\ (368N/mm^2)$$

(b) the total force on the pressure pad is equal to:

$$P = pF_{sl} = 36.8 \cdot 65\ 846 = 2\ 420\ 000kg = 2420\ tonnes\ (24MN)$$

Thus $\frac{P}{[P_{pr}]} = \frac{2420}{3000} \approx 0.81$, and this is entirely permissible.

11. In order to determine the piercing force, we find the length of the region of shear using equation (17), for $p_{podp} = 0$; $\tau_{sr} = 0.6 \cdot 1.8 = 1.08kg/mm^2$ $(10.8N/mm^2)$; the values of the coefficients a and b are found from Table 4.

$$\ell_{sr} = \frac{\sigma_0 + p_{podp}}{\tau_{sr}} \cdot \frac{0.5d_{ig}}{\frac{D_{tr}}{d_{ig}} + 1} = \frac{1.8(10-0.025 \cdot 100)}{1.08} \cdot$$

$$\frac{0.5 \cdot 100}{\frac{128}{100} + 1} = \frac{13.5 \cdot 50}{1.08 \cdot 2.28} = 274mm,$$

hence with $a = 0.08$ (see Table 5), equation (20) gives us the stress in the mandrel:

$$\sigma_{ig} = \sigma_s \left[10-0.025d_{ig} + \frac{4a(L_{sl} - \ell_{sr}) \cdot D_k}{(D_k-d_{ig})\ d_{ig}}\right] + p_{podp} =$$

$$1.8 \left[10-0.025 \cdot 100 + \frac{4 \cdot 0.08(288-274)306}{206 \cdot 100}\right] =$$

$$13.55kg/mm^2\ (135.5N/mm^2)$$

The piercing force is:

$$P_{prosh} = \sigma_{ig} \cdot F_{ig} = 13.55 \cdot \frac{\pi \cdot 100^2}{4} = 106\ 500kg =$$

$$106.5\ tonnes\ (10.65MN)$$

12. We find the maximum billet length on the basis of mandrel stability:

$$L_{sl} = 1.12\ d_{ig}\sqrt{\frac{E}{\sigma_s\ [10-0.025\ d_{ig}] \cdot n_u}} - H_{p.sh}$$

The mandrel stability margin factor is taken as 1.5-2.0. Therefore with T = 400°C and the corresponding value of E = 200 · 200 · 10^2kg/mm^2 (2 · 10^5N/mm^2):

$$L_{sl} = 1.12 \cdot 100\sqrt{\frac{200 \cdot 10^2}{1.8 \cdot 7.5 \cdot 1.8}} - 100 = 3100mm \gg 300mm;$$

i.e. the billet dimensions entirely satisfy the mandrel stability condition.

13. We determine the minimum permissible mandrel diameter on the basis of the tensile strength of the latter during extrusion (equation (33)).

We assume that, if the container is not heated (at the start of extruding a batch of tubes), there will be turbulent flow owing to the contact zones in the billet being greatly cooled. The length of the zone of slip along the mandrel (equation (29)) will therefore be:

$$\ell_{sk} = 0.8L_{sl} + \ell_{p.m} = 0.8 \cdot 303 + 10 = 252.4mm$$

Here the billet length L_{sl} is taken as being equal to the length found after piercing. The strength margin factor is taken as 1.5, allowing for possible changes in the tensile strength of the mandrel as its temperature rises from 400 to 600°C:

$$k = \frac{\sigma_b^{400}}{\sigma_b^{600}} \cdot 1.5 = \frac{135}{85} \cdot 1.5 = 2.38$$

For d_{ig} = 100mm, Table 8 gives us the value of the coefficient $c = \frac{\tau_{tr}}{\sigma_s} = 0.8;$

$$d_{ig} \geqslant \frac{4 \cdot 0.8 \cdot 252.4}{\frac{135}{1.8 \cdot 2.38} - 0.6\,(1 + \frac{36.8}{1.8})} = 46.5mm < 100mm$$

The mandrel therefore has a sufficient margin of tensile strength during extrusion.

14. We next calculate the piercing and extrusion time. With a hydraulic fluid pressure p_{akk} = 200atm (20MN/m^2) and a coefficient ξ_{ig} = 5.83 (see Chapter III), we find the piercing speed v_{pr}. With P_{prosh} = 106.5 tonnes (10.65MN), which corresponds to a fluid pressure in the mandrel cylinder of

$$p_{prosh} = \frac{P_{prosh}\,(kg)}{S_{ig}\,(cm^2)} = \frac{106\ 500}{2040} = 52.2 \text{ atm (bar)}$$

$$v_{prosh} = \sqrt{\frac{p_{akk} - p_{prosh}}{\xi_{ig}}} = \sqrt{\frac{200 - 52.2}{5.83}} = 84mm/s = 5.05m/min$$

Assuming a mean piercing speed v_{prosh} = 84mm/s, we now find the piercing time:

$$\tau_{prosh} = \frac{L_{sl}}{v_{prosh}} = \frac{288}{84} = 3.43s$$

We use equations (39) and (40) to determine the extruding speed and extrusion time.

The reduced extruding pressure:

$$p_{us} = \frac{P_{pr}}{S_{g.p.} + S_{ig}} = \frac{2\ 420\ 000}{12\ 300 + 2040} = 168.8 \text{ atm (bar)}$$

Hence, for ξ = 9.28 (Chapter II), we find the extruding speed:

$$v_{pr} = \sqrt{\frac{p_{akk} - p_{us}}{\xi}} = \sqrt{\frac{200 - 168.8}{9.28}} = 30.5mm/s = 1.83m/min$$

The extrusion time:

$$\tau_{pr} = \frac{L_{sl} - \ell_{ost}}{v_{pr}} = \frac{303 - 36.7}{30.5} = 8.75s$$

The total deformation time is:

$\tau_\Sigma = \tau_{prosh} + \tau_{pr} = 3.43 + 8.75 = 12.18s$

In this case, in order to avoid considerable cooling and under-extrusion τ_Σ must be lower than the values given in Table 25.

Table 25

Permissible time τ_Σ during the extrusion of tubes with with container not preheated s

Metal	$\frac{d_{ig}}{D_k}$					
	0.10	0.20	0.30	0.40	0.50	0.60
Copper and copper-nickel alloys	24*	22	20	18	16	14
Brasses, bronzes	34	32	30	28	26	24

* The values given for τ_Σ are for a steady extrusion rhythm after the container and the tool have been heated. When extrusion is carried out with the container not heated, the values of τ_Σ should be reduced by 30-50%.

CHAPTER II - DRAWING

Drawing is the most extensively used method for the cold deformation of metal. Various tube drawing processes have been described in references (14-16). Up to 80-85% of the total number of tubes produced by drawing are however produced with short or self-setting ("floating") mandrels, or without mandrels. In most cases finished tubes and tube blanks with diameters over 16-20mm are deformed on mandrels reducing both the diameter and the wall thickness simultaneously, while tubes with smaller diameters are deformed without a mandrel. In the latter case the diameter is reduced, and the alteration in wall thickness is governed by the parameters of the process (64). When small diameter tubes are produced with rigid wall thickness tolerances, they are often drawn down to a diameter of 3-5mm with a self-setting mandrel, or with a long mandrel-bar (15,16).

Tubes are drawn with short and floating mandrels, also without mandrels, on linear draw benches or blocks (14-16, 65-67) etc. The production schedules for drawing include the following operations:

1. pickling and washing of the blanks received from the heat treatment plant;
2. pushing on or ramming home the clamps;
3. the drawing operation itself;
4. annealing during the drawing of tubes made of metals and alloys subject to rapid work-hardening and loss of plastic properties;
5. straightening (in spiral roller machines, by stretching, etc);
6. annealing to give finished tubes the required mechanical properties;
7. cutting away the clamps and cutting into measured lengths.

The principal geometrical wastage in the drawing of tubes consists of the odd length ends and the discard containing the clamps. If the schedule is not correct (excessively high reduction per pass), if the tools and the mill are badly adjusted (the drawing axis and draw plate axis not coinciding, and high engagement speeds), or if the emulsion used as lubricant becomes contaminated or its composition alters, tubes have to be cut into sections, and this increases the consumption of metal.

Later we shall consider the problems governing the outputs of tube drawing plants, which consist of linear chain draw benches and blocks.

ANALYSIS OF THE RUNNING OF LINEAR TYPE CHAIN DRAW BENCHES

Determination of unutilised methods of raising output

The high outputs characteristic of modern draw benches are

achieved by means of multi-strand drawing, the mechanisation of operations such as applying tubes to the mandrels, feeding tubes into the draw plates, and removing drawn tubes, also high carriage movement speeds.

On linear chain type draw benches the tube drawing cycle time is made up of the basic (machine) time t_{mash} for actual drawing and the time consumed on auxiliary operations. In modern high-speed multi-strand draw benches the drawing cycle (70) includes the following operations (elements):

1. Feeding tubes into the "tribe" apparatus (time t_{pod}).
2. Applying the ends of tubes to the mandrels (t_{nad}).
3. Fitting the tubes with the tribe apparatus onto the mandrels and bars ($t_{t.a}$).
4. Adjusting the tubes on the mandrels (t_{pr}).
5. Tilting the rear table (t_{opr}).
6. Feeding tubes into the draw plates by push rods (t_{zad}).
7. Bringing the carriage up to the draw plate holder and gripping the tubes (t_{zakh}).
8. Drawing (t_{vol}).
9. Withdrawal of the carriage by inertia after drawing (t_{otk}).
10. Removal of drawn tubes onto the rack (t_{sb}).
11. Return of the carriage (t_{voz}).

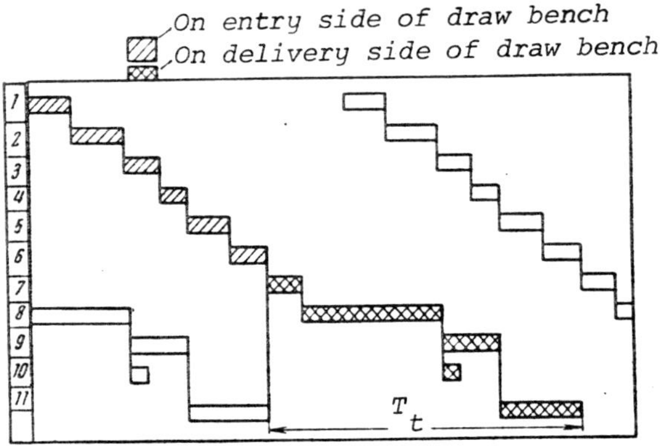

Fig.52: Cyclogram for tube draw bench

OPERATION ELEMENTS

1. Tubes fed into sockets of "tribe" apparatus
2. Ends of tubes applied to mandrels
3. Tubes fitted over bars and mandrels
4. Adjustment of tube positions
5. Tilting of rear table
6. Tubes fed into draw plates
7. Tubes gripped
8. Drawing
9. Travel of carriage
10. Removal of tubes
11. Return of carriage

Fig.52 is a cyclogram of the operation of a draw bench of this type. Elements 1-6 are at the entry side of the draw bench equipment, elements 7-11 at the delivery side. To attain high output figures with draw benches of this type it is necessary that the time for carrying out the preparatory elements of the drawing operation at the entry side of the draw bench (T_{vkh}) should be overlapped by the time taken on the elements at the delivery side (T_{vykh}).

In cases in which the operations at the entry side of the draw bench are overlapped by the operations at its delivery side, i.e. with:

$$T_{vkh} \leqslant T_{vykh} \tag{83}$$

the drawing tempo, and consequently also draw bench output, are governed by the period T_{vykh}.

Let us determine the minimum tube length L_{min} with which condition (83) is fulfilled. In accordance with what has been stated above,

$$T_{vkh} = t_{pod} + t_{nad} + t_{t.a.} + t_{pr} + t_{opr} + t_{zad} \tag{84}$$

and

$$T_{vykh} = t_{zakh} + t_{vol} + t_{otk} + t_{voz} \tag{85}$$

Equation (85) does not allow for the time t_{sb} for removing tubes and placing them on the rack, since this operation is usually carried out while the carriage is travelling.

The times taken for the individual elements of the process on the delivery side of the draw bench can be determined using the following equations:

(a) the drawing time:

$$t_{vol} = \frac{L_{tr} - \ell_r}{v_v} + \frac{\ell_r}{v'_v} \tag{86}$$

where:

ℓ_{tr} = the length of tube drawn in m;

ℓ_r = the carriage acceleration travel at the beginning of drawing in m;

v_v = the steady drawing process speed, in m/min;

v'_v = the mean acceleration period speed of the carriage, in m/min.

(b) the travel time:

$$t_{otk} = \frac{\ell_{otk}}{v_{sr.otk}} \tag{87}$$

where:

ℓ_{otk} = the inertia travel distance (in m) in which the speed of carriage movement slows from v_v (in m/min) to $v_v = 0$;

$v_{sr.otk}$ = the mean travel speed, $v_{sr.otk} = 0.5\, v_v$, with a linear law of reduction in speed;

(c) the carriage return time:

$$t_{voz} = \frac{L_{tr} + \ell_{otk} - \ell_t}{v_{voz}} + \frac{\ell_t}{v'_{voz}} \qquad (88)$$

where:

v_{voz} = the speed at which the carriage returns m/min;

v'_{voz} = the retarded speed of the carriage on approaching the draw plate holder in m/min;

ℓ_t = the return travel (in m) at the slower speed v'_{voz}.

The time taken to fit the tubes in the "tribe" apparatus onto the mandrel bars (at the entry side of a draw bench) can be determined using the equation:

$$t_{t.a} = \frac{L_{zag}}{v_{t.a}} = \frac{L_{tr}}{\mu \cdot v_{t.a.}} \qquad (89)$$

where:

L_{zag} = the length of the blank to be drawn, m;

μ = the reduction coefficient;

$v_{t.a}$ = the speed of the tube when fitted to the "tribe" apparatus bar, m/min.

The times taken to carry out the remaining elements of the process at the entry and delivery sides of the mill, which does not depend on the dimensions of the tube to be drawn, can be taken from the established norms.

If we substitute the values of T_{vkh} and t_{vykh} from equations (84) and (85), in equation (83), and find the values of t_{vol}, t_{voz} and $t_{t.a}$ using equations (87), (88) and (89) we get the equation:

$$t_{pod} + t_{nad} + \frac{L_{min}}{\mu \cdot v_{t.a}} + t_{pr} + t_{opr} + t_{zad} = t_{zakh} + \frac{L_{min} - \ell_r}{v_v} + \frac{\ell_r}{v'_v} + \frac{\ell_{otk}}{0.5 v_v} + \frac{L_{min} + \ell_{otk} - \ell_t}{v_{voz}} + \frac{\ell_t}{v'_{voz}} \qquad (90)$$

If we solve equation (90) for L_{min}, we get, in general form:

$$L_{min} = \frac{(t_{pod} + t_{nad} + t_{pr} + t_{opr} + t_{zad} + t_{zakh}) - \ell_r \left(\frac{v_V - v'_V}{v_V \cdot v'_V}\right)}{\left(\frac{1}{v_V} + \frac{1}{v_{voz}} - \frac{1}{\mu \cdot v_{t.a}}\right)} -$$

$$\frac{\frac{\ell_t}{v'_{voz}} + \frac{\ell_{otk}}{0.5 v_V} + \frac{\ell_{otk} - \ell_t}{v_{voz}}}{\left(\frac{1}{v_V} + \frac{1}{v_{voz}} - \frac{1}{\mu \cdot v_{t.a.}}\right)} \qquad (91)$$

As an example, we will find the minimum length of tube L_{min} for drawing on an automatic "Spidem" tube draw bench of the type fitted at a number of Soviet non-ferrous metal works (68-70).

The technical particulars of the "Spidem" draw bench are as follows:

Useful draw bench chain length L_{st} m	15
Drawing speed v_V m/min	5-45
Carriage return speed v_{voz}	100
Carriage return speed over last 2m of travel v'_{voz} m/min	46.2
Speed of tube fed onto tribe apparatus mandrel $v_{t.a}$ m/min	320
Initial drawing speed for 0.8-1.3m of travel v'_{voz} m/min	26.4
Inertia travel of carriage after end of drawing tube ℓ_{otk} m	2-5
Range of tubes mm:	
diameter	5-30
wall thickness	0.5-2.5

The times for the operation can be taken from the norms worked out for the "Spidem" draw bench by the "Krasnyi Vyborzhets" Works in Leningrad (min):

1 ..... t_{pod} = 0.0816	5 ..... t_{opr} = 0.0708
2 ..... t_{nad} = 0.0854	6 ..... t_{zad} = 0.0667
4 ..... t_{pr} = 0.0463	7 ..... t_{sbr} = 0.0502

On the basis of the technical particulars of the draw bench, we used the following figures: v_V = 45m/min; v'_V = 26.4m/min; v_{voz} = 100m/min; v'_{voz} = 46.2m/min; $v_{t.a}$ = 320m/min; ℓ_t = 2m; ℓ_{otk} = 3.0m; ℓ_r = 1.0m.

When the forces of friction slow the carriage down, the inertia travel time is determined using equation (87):

$$t_{otk} = \frac{3.0}{0.5 \cdot 45} = 0.1332 \text{ min}$$

This time greatly exceeds the tube removal time t_{sbr} = 0.0274 min.

Consequently during the natural travel of the carriage the element "removal of tubes" is entirely covered by the element "travel of carriage", and this provides grounds for removing t_{sbr} from T_{vykh} in equation (85).

We use equation (91) for determining the tube length for the case $T_{vkh} = T_{vykh}$ and $\mu = 1.3$:

$$L_{min} = \frac{(0.0816+0.0854+0.0463+0.0708+0.0667-0.502) - 1\left(\frac{45-26.4}{45\cdot 26.4}\right)}{\frac{1}{45}+\frac{1}{100}-\frac{1}{1.3\cdot 320}} -$$

$$\frac{\frac{3-2}{100}+\frac{2}{46.2}+\frac{3}{0.5\cdot 45}}{\frac{1}{45}+\frac{1}{100}-\frac{1}{1.3\cdot 320}} = 3.3m$$

The full overlapping of all the process elements at the entry side of a draw bench by the elements performed on the delivery side will thus only occur, with the maximum drawing speed (45m/min), with a drawn tube length $L_{tr} \geqslant 3.3$m. Figure 53 shows the relationship of the time T, the output in metres (M) and the output in tubes (N) to the length of tube drawn. When smaller length tubes are drawn the cycle time will be governed not by T_{vykh}, but by T_{vkh}, which is almost constant (only $t_{t.a}$ alters) in magnitude. This provides an approximately equal drawn bench output by number of tubes:

$$N = \frac{60}{T_{vkh}} \text{ tubes/h} \tag{92}$$

with any tube length (region I in fig.53).

When tubes with a length $L_{tr} > L_{min}$ are drawn, draw bench output is governed by the time T_{vykh}, equal to:

$$T_{vykh} = t_{zakh} + \underbrace{\frac{L_{tr}-\ell_r}{v_v} + \frac{\ell_r}{v'_v}}_{t_{vol}} + \underbrace{\frac{\ell_{otk}}{v_{sr.otk}}}_{t_{otk}} + \underbrace{\frac{L_{tr}}{v_{voz}} + \frac{\ell_{otk}-\ell_t}{v_{voz}} + \frac{\ell_t}{v'_{voz}}}_{t_{voz}} \tag{93}$$

With the present draw bench technical characteristics, i.e. with the values of v_v, v'_v, v_{voz} v'_{voz}, ℓ_r and ℓ_t constant, also with t_{zakh} = constant, an increase in output per pass defined as:

$$M = \frac{60}{T_{vykh}} \cdot L_{tr} \text{ m/min} \tag{94}$$

can only be attained by varying the variable quantities otk, $v_{sr.otk}$ and L_{tr}. If we say that $t_{vol} = t_{mash}$ and $t_{zakh} + t_{otk} + t_{voz} = t_{vs}$, it can be noted that the carriage inertia running time t_{otk}, for tubes with a length equal to $L_{tr} \leqslant L_{st} - \ell_{otk}$ is about 50% of the entire auxiliary time per pass.

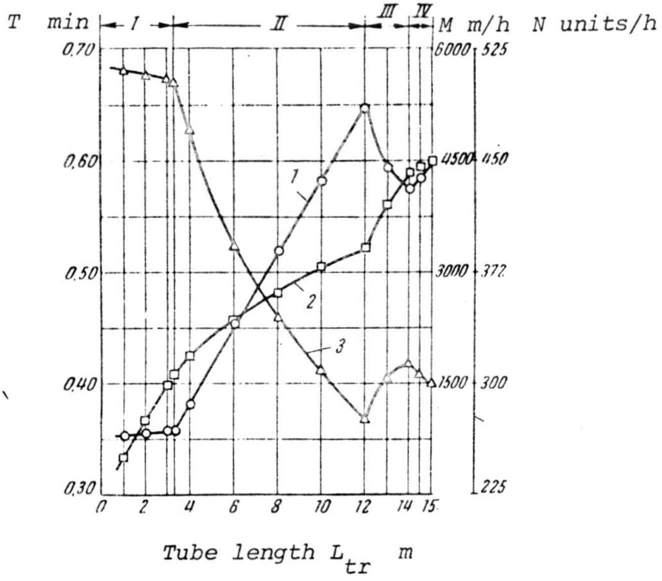

Fig.53: Relationship of the drawing time T_t (1) and output (2) metres and (3) production units an hour, to the length of tube drawn L_{tr} by three-strand drawing on "Spidem" draw bench.

When new draw benches are designed, to shorten the inertia travel ℓ_{otk} there must be an automatic system for slowing the carriage at the end of drawing a tube, and powerful enough end buffer stops.

As tube length L_{tr} is increased, draw bench output also increases (region II in fig.53), even if t_{otk} remains unchanged, owing to the smaller proportion occupied by the auxiliary time t_{vsp} in the total bench pass time ($t_{\Sigma} = t_{vsp} + t_{mash} = T_{vykh}$). This is due to the absolute increase in the machine time. The rate of increase in output rises as the carriage inertia travel is reduced (region III in fig.53), owing to the reduction in t_{otk}. In this case the carriage is stopped by striking the stops at the end of the chain. The drawing time is found by the equation:

$$t'_{otk} = \frac{\ell'_{otk}}{0.5\ (v_V + v_{ost})} = \frac{L_{st} - L_{tr}}{v_V \left(1 - \frac{L_{st} - L_{tr}}{2\ell_{otk}}\right)} \qquad (95)$$

It should be noted that ℓ_{otk} and t_{otk} should be reduced until the following condition is satisfied:

$$t_{otk} \geqslant t_{sbr}$$

In cases in which the time t_{otk} is overlapped by the tube removal time t_{sbr}, equation (85) will include t_{sbr} as the summand instead of t_{otk}. In cases in which $t_{sbr} \leqslant t'_{otk}$, the rate of increase in output drops abruptly (region IV in fig.53).

On the basis of the above considerations we are able to determine the rational limiting increase in drawn tube length starting from the element time t_{sbr}. With $t'_{otk} = t_{sbr}$, equation (95) gives us:

$$t_{sbr} = \frac{L_{st} - L_{max}}{v_V \left(1 - \frac{L_{st} - L_{max}}{2\ell_{otk}}\right)}$$

hence:

$$L_{max} = L_{st} - \frac{1}{\frac{1}{v_V \cdot t_{sbr}} + \frac{1}{2\ell_{otk}}} \qquad (96)$$

In the base of the "Spidem" draw bench, with v_V = 45m/min, ℓ_{otk} = 3m and t_{sbr} = 0.0274:

$$L_{max} = 15 - \frac{1}{\frac{1}{45 \cdot 0.0274} + \frac{1}{2 \cdot 3}} \approx 14m$$

Equivalent calculations can be made for chain draw benches cf different designs, also when new draw benches are being designed.

CALCULATING CHAIN DRAW BENCH OUTPUT

The total time taken to draw a tube in k passes is equal to:

$$t_{\Sigma} = \sum_{i=1}^{k} (t_{mash} + t_{vsp})_i = \sum_{i=1}^{k} t_{mash(i)} + \sum_{i=1}^{k} t_{vsp(i)} \qquad (97)$$

We now find the total machine time (by A M Rytikov's method).

For the k-th (last) pass:

$$t_{mash(k)} = \frac{L_{tr} - \ell_r}{v_v} + \frac{\ell_r}{v'_v}$$

<u>Calculating chain draw bench output</u>

For the (k-1)-th pass:

$$t_{mash(k-1)} = \frac{\frac{L_{tr}}{\mu_k} - \ell_r}{v_v} + \frac{\ell_r}{v'_v}$$

For the (k-2)-th pass:

$$t_{mash(k-2)} = \frac{\frac{L_{tr}}{\mu_k \cdot \mu_{k-1}} - \ell_r}{v_v} + \frac{\ell_r}{v'_v}$$

. . . .

For the 2nd pass:

$$t_{mash(2)} = \frac{\frac{L_{tr}}{\mu_k \cdot \mu_{k-1} \cdots \mu_4 \cdot \mu_3} - \ell_r}{v_v} = \frac{\ell_r}{v'_v}$$

For the 1st pass:

$$t_{mash(1)} = \frac{\frac{L_{tr}}{\mu_k \cdot \mu_{k-1} \cdots \mu_4 \cdot \mu_3 \cdot \mu_2} - \ell_r}{v_v} + \frac{\ell_r}{v'_v}$$

If we total t_{mash} and assume:

$$\mu_1 = \mu_2 = \ldots = \mu_{k-1} = \mu_k = \mu_{sr} = \sqrt[k]{\mu_1 \cdot \mu_2 \cdots \mu_k} \qquad (98)$$

we find that:

$$\sum_{i=1}^{k} t_{mash(i)} = \frac{L_{tr}\left(1 + \frac{1}{\mu_{sr}} + \frac{1}{\mu_{sr}^2} + \ldots + \frac{1}{\mu_{sr}^{k-1}}\right) - k\ell_r}{v_v} + \qquad (99)$$

$$\frac{k\ell_r}{v'_v} = L_{tr} \cdot \frac{\mu_{sr}^k - 1}{(\mu_{sr}-1)\mu_{sr}^{k-1}} \cdot \frac{1}{v_v} + k\ell_r\left(\frac{1}{v'_v} - \frac{1}{v_v}\right)$$

In a similar manner we find the total auxiliary time for k passes. The auxiliary time for the k-th pass is equal to:

$$t_{vsp(k)} = \frac{\ell_{otk}}{0.5v_V} + \frac{L_{tr}}{v_{voz}} + \frac{\ell_{otk}-\ell_t}{v_{voz}} + \frac{\ell_t}{v'_{voz}} = \frac{L_{tr}}{v_{voz}} + T$$

Obviously with ℓ_{otk} and ℓ_t constant, the sum:

$$\frac{\ell_{otk}}{0.5v_V} + \frac{\ell_{otk} - \ell_t}{v_{voz}} + \frac{\ell_t}{v'_{voz}} = T$$

is the same in all the passes.

For the $(k-1)$-th pass:

$$t_{vsp(k-1)} = \frac{L_{tr}}{\mu_k \cdot v_{voz}} + T$$

For the $(k-2)$-th pass:

$$t_{vsp(k-2)} = \frac{L_{tr}}{\mu_k \cdot \mu_{k-1} \cdot v_V} + T$$

. . . .

For the 2nd pass:

$$t_{vsp(2)} = \frac{L_{tr}}{\mu_k \cdot \mu_{k-1} \cdots \mu_4 \cdot \mu_3 \cdot v_V} + T$$

For the 1st pass:

$$t_{vsp(1)} = \frac{L_{tr}}{\mu_k \cdot \mu_{k-1} \cdots \mu_4 \cdot \mu_3 \cdot \mu_2 \cdot v_V} + T$$

If we total t_{vsp} from the 1st to the k-th pass inclusive, and assume condition (98), we get:

$$\sum_{i=1}^{k} t_{vsp(i)} = \frac{L_{tr}\left(1 + \frac{1}{\mu_{sr}} + \frac{1}{\mu_{sr}^2} + \ldots + \frac{1}{\mu_{sr}^{k-1}}\right)}{v_V} + kT =$$

$$L_{tr} \frac{\mu_{sr}^k - 1}{(\mu_{sr} - 1)\mu_{sr}^{k-1}} \cdot \frac{1}{v_V} + kT \qquad (100)$$

The total time for k passes is equal to:

$$t_{\Sigma} = \left(L_{tr} \left(\frac{1}{v_v} + \frac{1}{v_{voz}} \right) \cdot \frac{\mu_{sr}^{k} - 1}{(\mu_{sr} - 1)\mu_{sr}^{k-1}} \right) +$$

$$k\left(t_{zakh} + \ell_r \left(\frac{1}{v'_v} - \frac{1}{v_v} \right) + \frac{\ell_{otk} - \ell_t}{v_{voz}} + \frac{\ell_t}{v'_{voz}} + \frac{\ell_{otk}}{0.5v_v} \right) \qquad (101)$$

The total time t_{Σ} expended on producing one finished tube by drawing in k passes enables us to determine the mill output an hour (M) for single-strand drawing:

$$M = \frac{60}{t_{\Sigma}} \cdot L_{tr} \text{ (in metres of tube drawn)} \qquad (102)$$

also for three-strand drawing:

$$M = \frac{60}{t_{\Sigma}} \cdot 3L_{tr} \qquad (102a)$$

Using the nomograms in fig.54 it is easy approximately* to calculate (a) the time t_{Σ}, and (b) the initial billet length. (*Equation (101) provides an accuracte solution with $L_{tr} \leqslant L_{st} - \ell_{otk}$.)

Drawing with intermediate shearing

As the length of tube drawn is increased, draw bench output rises because the proportion occupied by the auxiliary time t_{vsp} in the total time decreases. When schedules for drawing tubes are being compiled, therefore, attempts must be made to utilise the useful draw bench length to the full. This can most effectively be done if tubes are drawn using multiple length billets and there is an operation of shearing after one of the passes. This type of schedule makes it possible to reduce the total number of passes necessary to produce the same number of tubes. This can be illustrated as follows. The number of passes necessary to produce x tubes from single length billets by single-strand drawing is k, and consequently:

$$\sum_{i=1}^{k} t_{vsp(i)} = x \cdot kt_{vsp}$$

but if a multiple length billet is used and cut into two parts after the i-th pass, to produce the same number of tubes requires

$$\sum_{i=1}^{k} t_{vsp(i)} = \frac{x}{2} \cdot i \cdot t_{vsp} + x(k-i)\, t_{vsp} = xt_{vsp} \left(k \frac{i}{2} \right)$$

The amount of time saved will be $x \cdot t_{vsp} \cdot \frac{i}{2}$

We shall prove later that the use of a large number of intermediate shearing operations (for instance after passes i and j) is not an economic proposition, since the expenditure on shearing and on transporting the tubes to the saws and from them back to the draw bench increases. Moreover, numerous transportations may in turn increase the amount of tubes rejected for dents, etc.

It should be noted that the machine time $\sum_{i=1}^{k} t_{mash(i)}$ and the total number of draws are the same whether or not there is an intermediate shearing operation. When drawing is carried out without shearing, it takes x drawing operations to produce x tubes (in cases in which the technology does not provide for an intermediate shearing operation and a second draw). When a double length billet is drawn and cut into two parts after pass i there are $\frac{x}{2}$ drawing operations before the first pass and a further $\frac{x}{2}$ drawing operations after shearing and before pass $(i+1)-M$. The total number of drawing operations is therefore $\frac{x}{2} + \frac{x}{2} = x$. We assume that the initially forged grips are not cut away and can be used at the second ends of the tubes.

Let us determine the time saved per tube when tubes are drawn in n strands from the first to the last, k-th pass, and cut into $\frac{m_1}{m_2}$ parts after pass i, then into $\frac{m_2}{m_3}$ parts after pass j and into m_3 parts after the final, k-th pass:

	No cutting	With cutting
Number of passes	k	k
Number of stands	n	n
Multiplying factors:		
from pass 1 to pass i	1	m_1
from pass $(i + 1)$ to pass j	1	m_2
from pass $(j + 1)$ to pass k	1	m_3
Auxiliary time per pass per tube:		
from pass 1 to pass i	$\frac{t_{vsp}}{n}$	$\frac{t_{vsp}}{nm_1}$
from pass $(i + 1)$ to pass j	$\frac{t_{vsp}}{n}$	$\frac{t_{vsp}}{nm_2}$

	No cutting	With cutting
from pass (*j* + *1*) to pass *k*	$\frac{t_{vsp}}{n}$	$\frac{t_{vsp}}{nm_3}$
Total auxiliary time $\sum_{i=1}^{k}$ per tube:		
for *i* passes	$\frac{t_{vsp}}{n} i$	$\frac{t_{vsp}}{nm_1} i$
for (*j* - *1*) passes	$\frac{t_{vsp}}{n}(j-i)$	$\frac{t_{vsp}}{nm_2}(j-i)$
for (*k* - *j*) passes	$\frac{t_{vsp}}{n}(k-j)$	$\frac{t_{vsp}}{nm_3}(k-j)$
from pass *1* to pass *k*	$\frac{t_{vsp}}{n}i+\frac{t_{vsp}}{n}(j-i)+\frac{t_{vsp}}{n}(k-j) = \frac{t_{vsp}}{m}k$	$\frac{t_{vsp}}{nm_1}i+\frac{t_{vsp}}{nm_2} x (j-1)+\frac{t_{vsp}}{nm_3} x (k-j)=\frac{t_{vsp}}{n} x \left(\frac{i}{m_1}+\frac{j-i}{m_2}\frac{k-j}{m_3}\right)$

The time saved per tube drawn with cutting is:

$$\Delta t_{ek} = \frac{t_{vsp}}{n} k - \frac{t_{vsp}}{n}\left(\frac{i}{m_1}+\frac{j-i}{m_2}+\frac{k-j}{m_3}\right) = \frac{t_{vsp}}{n}\left(k-\frac{i}{m_1}-\frac{j-i}{m_2}-\frac{k-j}{m_3}\right) \qquad (103)$$

When long tubes are drawn, if the inequality $L_{tr}>L_{st}-\ell_{otk}$ is valid, the true value of Δt_{ek} will be greater than the value calculated using equation (103).

Example 1: We wish to find the time saved per tube when tubes are drawn in five passes in three strands with intermediate cutting in half after passes 2 and 4, by comparison with drawing without cutting.

Pass	Drawing procedure	Number of strands	Multiplication factor for pass	Saving of t_{vsp}
1	A 	3	4	$\frac{3}{4.3}$
2	B 	3	4	$\frac{3}{4.3}$
3	C 	3	2	$\frac{1}{2.3}$
4	D 	3	2	$\frac{1}{2.3}$
5	E 	3	1	0

According to equation (103), with $i = 2$; $j = 4$; $k = 5$; $n = 3$; $m_1 = 4$; $m_2 = 2$; $m_3 = 1$, we get:

$$\Delta t_{ek} = \frac{t_{vsp}}{3} \left(5 - \frac{2}{2} - \frac{4-2}{2} - \frac{5-4}{1}\right) = \frac{t_{vsp}}{3} \frac{5}{2} = \frac{5}{6} t_{vsp}$$

<u>Example 2:</u> We wish to find the saving of time per tube in the case of drawing in one strand in the 1st pass and three strands in the subsequent three passes, with the tubes cut in half after the 1st and 3rd passes.

Pass	Drawing procedure	Number of strands	Multiplication factor for pass	Saving of t_{vsp}
1	B 	1	4	$\frac{1}{4.1}$
2	C 	3	2	$\frac{1}{2.3}$
3	D 	3	2	$\frac{1}{2.3}$
4	E 	3	1	0

In order to solve the problem using equation (103), it is subdivided into two parts. For part 1: $i_1 = j_1 = k_1 = 1$; $n_1 = 1$; $m_1 = m_2 = m_3 = 4$. For part 2: $i_2 = j_2 = 2$; $k_2 = 3$; $n_2 = 3$; $m_1 = m_2 = 2$; $m_3 = 1$.

$$t_{ek} = \frac{t_{vsp}}{1}\left(1 - \frac{1}{4}\right) + \frac{t_{vsp}}{3}\left(3 - \frac{2}{2} - \frac{3-2}{1}\right) = \frac{13}{12}\; t_{vsp}$$

The value of Δt_{ek} enables us to gauge the economic advisability of using procedures with intermediate shearing. In cases in which the time norm t_{rez} for cutting one tube is shorter than Δt_{ek} (i.e. $t_{rez} < \Delta t_{ek}$), the tube production cycle is shortened.

Let us, as an example, consider different methods of producing 10m long finished tubes in five passes with a mean reduction $\mu = 1.3$ per pass. With a useful chain length $L_{sg} = 15$m, there are 9 different possible technologies of which 8 are practical possibilities:

I - drawing with no intermediate shearing;
II - with the tube cut in half after the 4th pass;
III - with the tube cut in half after the 3rd pass;
IV - with the tube cut in half after the 2nd pass;
IVa - with the tube cut in three parts after the 2nd pass;
V - with the tube cut in half after the 1st pass;
Va - with the tube cut in four parts after the 1st pass;
Vb - with the tube cut in halves after the 1st and 3rd passes;
VI - with the tube cut into lengths of 2/3 L_2 and 1/3 L_2 after the 2nd pass, and with the longer tube cut in half after the 3rd pass.

For the calculations, we make the following assumptions:

(a) the engagement time $t_{zakh} = 0.05$min;
(b) the carriage inertia travel with natural retardation $\ell_{otk} = 5$m;
(c) the carriage travel at the retarded return speed $\ell_t = 2$m;
(d) the drawing speed $v_v = 42$m/min;
(e) the carriage acceleration travel at the beginning of drawing $\ell_r = 1$m;
(f) the carriage acceleration speed for the distance ℓ_r $v'_v = 25$m/min.
(g) the carriage return speed $v_{voz} = 100$m/min.

When output is being determined, the total time taken to draw one finished tube in k passes is found using the equation:

$$t_\Sigma = \sum_{i=1}^{k} \frac{t_{\Sigma(i)}}{m_i n_i} \qquad (104)$$

where:

m_i = the tube multiplication factor in pass i;
n_i = the number of strands used for drawing in pass i;
$t_{\Sigma(i)}$ = the sum of the machine and auxiliary times in pass i.

Table 26 Comparison of outputs from a "spidem" three-strand draw bend when 10m long tubes are drawn in five passes by different technological procedures (reduction per pass μ = 1.3)

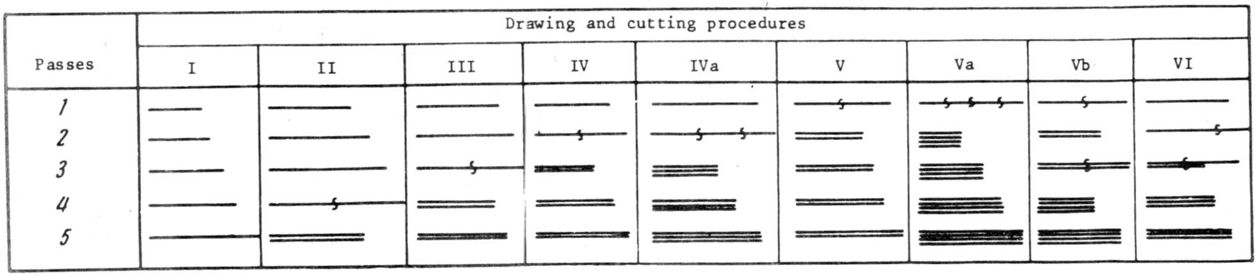

Process particulars	Procedure Nos.								
	I	II	III	IV	IVa	V	Va	Vb	VI
Length of one tube m:									
L_5	10,0	10,0	10,0	10,0	10,0	10,0	10,0	10,0	10,0
L_4	7,70	7,70×2 = = 15,40*	7,70	7,70	7,70	7,70	7,70	7,70	7,70
L_3	5,90	—	5,9×2 = = 11,8	5,90	5,90	5,90	5,90	5,9×2 = = 11,8	5,9×2 = = 11,8 : 2 = = 5,90
L_2	4,55	—	9,10	4,55×2 = = 9,10	4,55×3 = = 13,65	4,55	4,55	9,10	9,1+4,55 = = 13,65
L_1	3,50	—	7,0	7,0	10,5	3,50×2=7,0	3,5×4 = 14	7,0×2 = 14	10,5
L_{zag}	2,69	—	5,38	5,38	8,07	5,38	10,76	10,76	8,07
Number of finished tubes per blank	1	—	2	2	3	2	4	4	3
Total time for drawing n tubes min	2,966	—	4,648	5,176	7,099	5,553	10,447	8,667	6,568
Total time per finished tube min	2,966	—	2,324	2,588	2,366	2,777	2,612	2,167	2,187
Drawing output m/h:									
One strand	202	—	258	232	254	216	229	277	274
Three strand	606	—	774	696	762	648	687	831	822

* This variation is impossible since the tube is longer than the draw bench

Table 26 gives the results of calculations made for drawing by the different procedures. The procedure for drawing tubes without cutting (I) which is much used also has the lower output. All the other procedures provide increased outputs: 6.9% with procedure V, about 25.8% with procedure IVa, 35.6% with procedure VI, and 37.1% with procedure Vb.

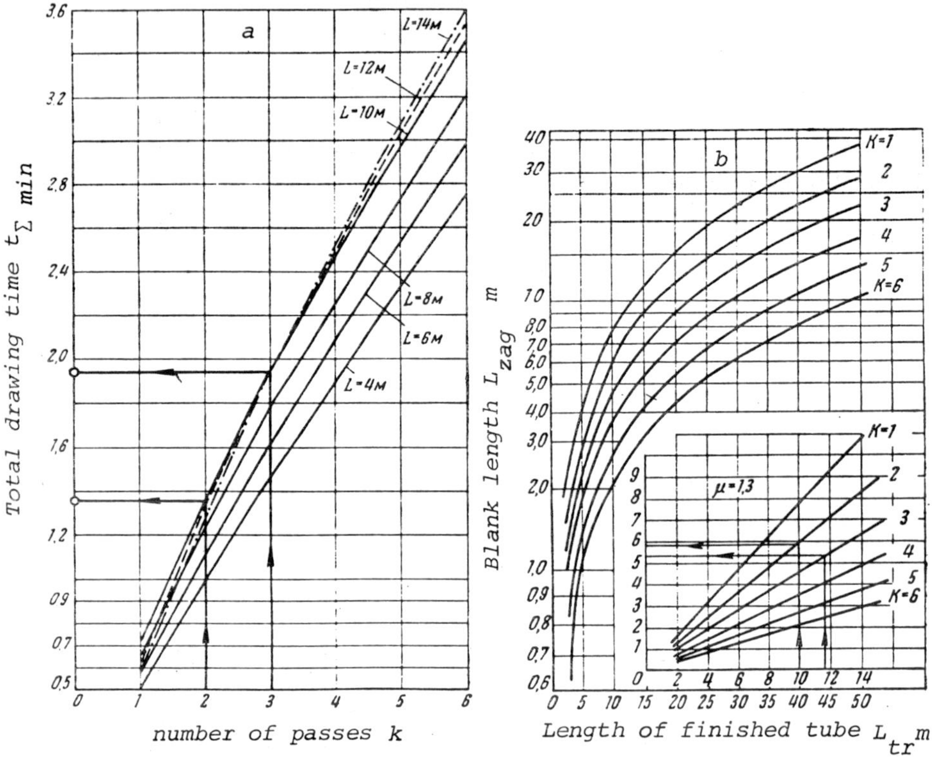

Fig.54: Nomograms for determining (a) the total drawing time t_Σ and (b) the blank length for different finished tube lengths L_{tr} and numbers of passes k; the following data were used in calculating the nomograms:

$\mu_{sr} = 1.3$; $v_{vol} = 42$m/min; $v'_{vol} = 25$m/min; $\ell_r = 1.0$m;

$\ell_T = 2.0$m; $v_{voz} = 100$m/min; $v'_{voz} = 45$m/min; $\ell_{otk} = 5$m;

useful draw length $L_{st} = 15$m

In order approximately to assess output and select the optimum procedure, the nomograms in fig.54 can be used. These nomograms give the example of calculating the drawing of a 10m long tube by procedure III, including cutting in half after pass 3.

1. We use the nomogram in fig.54b for determining the length of the blank after pass 3 for the subsequent drawing of tubes in the two last passes; L = 5.90m. Before shearing the tube, it must have double the length after pass 3, i.e. 5.90 x 2 = 11.80m (we will round it off to 12.0m). This length is less than the useful draw bench length L_{st} = 15m, and this indicates that this particular procedure is a practical possibility.

2. We determine the total time for drawing two 10m long tubes in the last two passes (t_Σ = 2 x 1.35 = 2.70 min), and for drawing one tube 12m long in the first three passes (t_Σ = 1.94 min) using the nomogram in fig.54a. The overall total time is: 1.94 + 2.70 = 4.64 min.

3. With single-strand drawing the output is:

$$M = \frac{60}{t_\Sigma} NL_{tr} = \frac{60}{4.64} \quad 2 \cdot 10 = 258 m/h$$

with three-strand drawing, it is 258 x 3 = 774m/h.

4. The absolute length of the initial blank can be determined, for μ_{sr} = 1.3, from the nomogram in fig.54b. As was stated earlier, for the procedure under consideration the length of the cut tube after pass 3 must be 5.9m, and before cutting the length of the tube must be 5.9 x 2 = 11.8m. Hence we find that the length of the initial blank (with k = 3) is ~5.4m. Output is calculated more accurately using equations (85) and (104).

When the procedure is selected, it is necessary to be guided by the productive capacity of the production equipment, also by considerations of technology. For instance the blanks must be transportable, and their length must not exceed the permissible length for processing with the existing equipment, for instance it must not be longer than the pickling baths. When the decision is taken on cutting, allowance must be made for the possibility for forging grips to the work-hardened tubes and the strength of these grips, which drops as the tube wall thickness is reduced.

When the procedure to use is selected, this must be done on the basis of raising the output from the main equipment (presses and draw·benches); this is more important to production than placing additional loads on the auxiliary equipment (saws and forging hammers).

ANALYSIS OF OPERATION OF DRAWING BLOCKS

Calculation of drawing block output

Drawing block output is determined using the equation:

$$M = \frac{60}{t_\Sigma} \ell_{tr} \ m/h \tag{105}$$

where:

t_{Σ} = the total drawing time consumed on producing a finished tube min;

ℓ_{tr} = the finished tube length m.

The time t_{Σ} for k passes is determined by the equation:

$$t = \sum_{i=1}^{k} t_{mash(i)} + \sum_{i=1}^{k} t_{vsp(i)} \qquad (106)$$

Here:

$t_{mash(i)}$ = the machine time for drawing in the i-th pass;

$t_{vsp(i)}$ = the auxiliary time for the i-th pass.

When tubes are drawn with a floating mandrel, the time t_{vsp} is consumed on the following operations (67):

1. Installation of coil;
2. Cutting off tube end;
3. Fitting of felt pad;
4. Spraying lubricant;
5. Positioning mandrel;
6. Forging tube end in a rotating forging machine;
7. Fitting forged end of tube into die;
8. Gripping;
9. Switching on drawing block.

When drawing is carried out without a mandrel, operations 3-5 are eliminated, and operations 2 and 6 can be replaced by adjustment of the grip.

The total time t_{Σ} can be calculated by summing up the times for drawing in the different passes:

$$t_{vol} = t_{mash} + t_{vsp}$$

The drawing time in the last, k-th, pass:

$$t_k = \frac{\ell_{tr}}{v} + t_{vsp}.$$

in the (k-1)-th pass:

$$t_{k-1} = \frac{\ell_{tr}}{\mu_k v} + t_{vsp}$$

in the (k-2)-th pass:

$$t_{k-2} = \frac{\ell_{tr}}{\mu_k \mu_{k-1} v} + t_{vsp}$$

. . . .

in the 2nd pass:

$$t_2 = \frac{\ell_{tr}}{\mu_k \mu_{k-1} \ldots \mu_4 \mu_3 v} + t_{vsp}$$

in the 1st pass:

$$t_1 = \frac{\ell_{tr}}{\mu_k \mu_{k-1} \ldots \mu_4 \mu_3 \mu_2 v} + t_{vsp}$$

Here v is the drawing speed (the circumferential velocity of the block) in m/min; $\mu_k, \mu_{k-1}; \ldots; \mu_2$ are the reductions by passes.

Totalling up the time by passes, we get:

$$t_\Sigma = t_k + t_{k-1} + \ldots + t_2 + t_1 = \frac{\ell_{tr}}{v}$$

$$\left(1 + \frac{1}{\mu_k} + \frac{1}{\mu_k \mu_{k-1}} + \ldots + \frac{1}{\mu_k \mu_{k-1} \ldots \mu_3 \mu_2} \right) + kt_{vsp}$$

In the last expression, k is the total number of passes.

We will assume that $\mu_k = \mu_{k-1} = \ldots = \mu_2 = \mu_1 = \mu$; the expression in parentheses then becomes:

$$1 + \frac{1}{\mu} + \frac{1}{\mu^2} + \ldots + \frac{1}{\mu^{k-1}} = \frac{\mu^k - 1}{\mu^{k-1}(\mu - 1)}$$

Hence the total drawing time is:

$$t_\Sigma = \frac{\ell_{tr}}{v} \frac{\mu^k - 1}{\mu^{k-1}(\mu - 1)} + kt_{vsp} \qquad (107)$$

Relationship of output to the main parameters of the process

We can see from equations (105) and (107) that block output in metres of finished tubes is governed by:

1. the length of finished tube ℓ_{tr} m;
2. the drawing speed v in m/min;
3. the mean reduction per pass μ;
4. the total number of passes k;
5. the auxiliary time t_{vsp} min.

Auxiliary time, mean reduction and number of passes. When long tubes are drawn, in most cases operations 1-6 are covered by the machine time. Therefore t_{vsp} is made up of the time taken to fit the tube in the die, grip it, and switch on the block (machine and manual time). In modern mechanised blocks, in each pass t_{vsp} = 0.5-0.75 min. It should be noted that, with a total number of passes $k > 9$ the magnitude of the multiple

$$\frac{\mu^k - 1}{(\mu - 1)\,\mu^{k-1}}$$

in equation (107) alters little. The total machine time Σt_{mash}, with $k > 9$, therefore scarcely alters, and the total time increases owing to changes in Σt_{vsp} (Table 27).

Table 27

Alteration in the relationship of Σt_{mash} and Σt_{vsp} as the total number of passes k is increased. Mean reduction per pass $\mu = 1.5$ length of finished tube $\ell_{tr} = 5000m$, drawing speed $v = 500m/min$, Auxiliary time $t_{vsp} = 1.0min$

k	$\frac{\mu^k - 1}{(\mu-1)\,\mu^{k-1}}$	Σt_{mash}	Σt_{vsp}	t_Σ	$\frac{\Sigma t_{vsp}}{t_\Sigma} \cdot 100\%$
3	2.11	21.1	3	24.1	12.5
6	2.74	27.4	6	33.4	18.0
9	2.92	29.2	9	38.2	23.6
12	2.98	29.8	12	41.8	28.7
15	3.00	30.0	15	45	33.4

We can see from the figures in Table 27 that, as the number of passes is increased, the proportion occupied by the auxiliary time in the total drawing time also increases. When we analyse equation (107), we can draw the conclusion that the proportion occupied by the auxiliary time decreases as tube length is increased and the drawing speed lowered.

If there are a large number of passes and a high drawing speed the auxiliary time may become the factor which governs block output. As the mean reduction μ is reduced, the number of passes k increases, and this results in a relative increase in the auxiliary time and reduction in output (Table 28).

Drawing speed. Increasing the drawing speed results in a rise in block output, a reduction in total time, and a relative increase in auxiliary time. Figure 55 contains curves for the relationship of block output ot the number of passes for speeds v = *100, 200, 400 and 800m/min* and different finished tube lengths. When tubes are drawn with a reduction μ = *1.3* and a length of *1800m* from blanks *50m* long with t_{vsp} = *0.5min and k = 14*, increasing the drawing speed by *2, 4, or 8* times from *100m/min* results in respective increases in output of *70, 220, and 400%*. The reason why the increases in speed and output are not proportional to one another is that the proportion occupied by auxiliary time in the total time becomes greater.

Table 28

Block output with relation to the Mean Reduction per pass. Total Reduction $\mu_{\Sigma} = 3.4$; $v = 50m/min$; $t_{vsp} = 0.5min$; $l_{tr} = 100m$; $l_{zag} = 29.5m$

Mean redn. per pass μ	No. of passes	$\frac{\Sigma t_{vsp}}{t_{\Sigma}} \cdot 100\%$	Output m/h	Increase in output %
1.28	5	29.1	700	
1.50	3	21.5	860	22.9

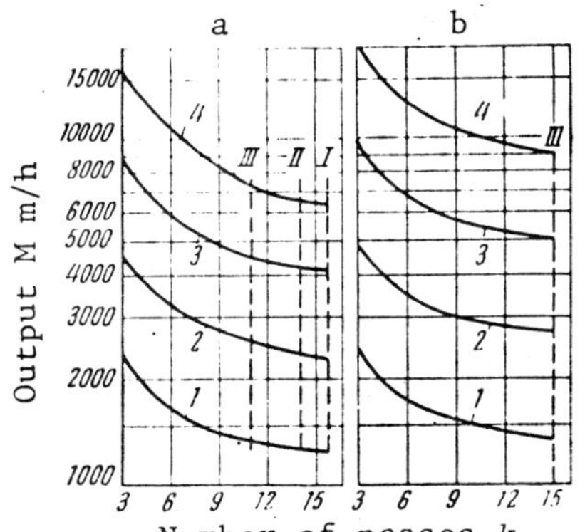

Fig.55: Relationship of drawing block output M, in metres of tube drawn, to the number of passes k, for different drawing speeds. Finished tube lengths: a) 1800m; b) 5400m; Initial blank lengths in m: I) 25m II) 50m; III) 100m; drawing speeds, in m/min: 1) 100; 2) 200; 3) 400; 4) 800. Mean reduction per pass μ = 1.3; auxiliary time t_{vsp} = 0.5 min.

Tube length. Equations (105) and (107) can be written in the following form:

$$M = \frac{60}{\frac{1}{v}\,\frac{\mu^k - 1}{\mu^{k-1}\,(\mu-1)} + \frac{kt_{vsp}}{\ell_{tr}}} \qquad (108)$$

The term $\frac{kt_{vsp}}{\ell_{tr}}$ in the denominator represents the proportion of the total auxiliary time applicable to unit length of tube. With k and t_{vsp} constant, as ℓ_{tr} increases the term decreases, and this indicates reduction in the proportion occupied by the auxiliary time. As tube length is increased the rate of increase in output decreases steadily. One would expect the greatest effect from increasing tube length with high drawing speeds (see the increases in output in Table 29 with $k = 3$ and $v = 200$ and $800 m/min$) and large numbers of passes (see increase in output Table 29 with $v = 200 m/min$ and $k = 3$ and 6).

Table 29

Drawing Block output with Different Numbers of Passes and Different Tube Lengths. Mean Reduction per pass $\mu = 1.5$; $t_{vsp} = 0.5 min$

No. of passes	Drawing speed m/min	Length of finished tube m	Output m/h	Increase in output %
3	200	200 1800	3330 5270	— 58.2
	400	200 1800	4 690 10 800	— 130
	800	200 1800	5 900 18 800	— 219
6	200	200 1800	2130 3900	— 83

RELATIONSHIP BETWEEN THE RUNNING OF DRAWING BLOCKS AND THE PLANT PRODUCING BLANKS

Tube blanks for block drawing are produced either on chain draw benches or in cold tube rolling mills (TCRM). To link the running of the plant producing the tube blanks and the drawing blocks it is sometimes advisable to represent drawing block hourly output in metres of blanks drawn. In this case the total time consumed on drawing the tube blank in *k* passes is also determined by means of equation (107).

The drawing times by passes are:

1st pass:

$$t_1 = \frac{\ell_{zag}\mu_1}{v} + t_{vsp}$$

2nd pass:

$$t_2 = \frac{\ell_{zag}\mu_1\mu_2}{v} + t_{vsp}$$

....

the k-th pass;

$$t_k = \frac{\ell_{zag}\mu_1\mu_2\cdots\mu_k}{v} + t_{vsp}$$

Then:

$$t_\Sigma = \frac{\ell_{zag}}{v}(\mu_1 + \mu_1\mu_2 + \ldots + \mu_1\mu_2\mu_3 \ldots \mu_k) + kt_{vsp}$$

Assuming $\mu_1 = \mu_2 = \ldots = \mu_k = \mu$, we get*

$$t_\Sigma = \frac{\ell_{zag}}{v} \; \frac{\mu(\mu^k - 1)}{\mu - 1} + kt_{vsp}. \tag{109}$$

* We obtain the same equation if we assume $\ell_{tr} = \ell_{zag}\cdot\mu_k$ and substitute it in equation (107).

The number of tubes drawn in an hour $N = \frac{60}{t_\Sigma}$

with a total number of passes $N \cdot k$

Block output in metres of blanks drawn an hour

$$M = \frac{60}{t_\Sigma} \ell_{zag} \tag{110}$$

Block output in *kg/h*

$$G = \frac{\gamma \pi D_{sr} s \ell_{zag} 60}{t_\Sigma}$$

where:

γ = the specific gravity of the metal of which the tubes are made

D_{sr} = the mean tube diameter, equal to $D_{tr} - s$;

s = the tube wall thickness

Let us, as an example, consider the technology in which, after tubes have been rolled in a TCRM, they are drawn with drawing blocks. For instance, in two drawing operations tubes of 9 x 1.0mm can be produced (reduction per pass 1.35) from 16 x 1.0mm tube blanks. Tubes with these dimensions can be produced in four passes (with the same reduction) from 28 x 1.0mm tube blanks. Both the first and the second blanks can be rolled in KhPT-32 mills with the same output in metres. With equal outputs in metres, when 16 x 1 and 28 x 1mm tubes are produced, in the latter case the weight of tubes rolled in the same period of time is 78% greater.

Hence the problem arises of linking up the outputs of the cold tube rolling mill and the drawing blocks and of deciding upon the economically rational dimensions of the tube blanks (semi-rolled products). Obviously the output of a cold tube rolling mill, in metres of rolled tubes, and the output of a block, in metres of blank required, must be equal to:

$$\frac{m\mu_\Sigma n}{1000} \eta TCRM = \frac{1}{\frac{1}{v} \frac{\mu(\mu^k - 1)}{\mu - 1} + \frac{kt_{vsp}}{\ell_{zag}}} \tag{111}$$

With a prescribed blank length ℓ_{zag} governed by the length of of the receiving table of the tube cold rolling mill (TCRM), a prescribed reduction μ per pass, a prescribed drawing speed v etc., the equation obtained may be solved for k, and consequently the rational dimensions of the tube blanks for the blocks can be determined.

When tubes 25m long are rolled, with a KhPT-32 mill output of 324m/h ($m \mu_{\Sigma}$ = *60mm, n = 100 strokes/min*) and a drawing speed of 45m/min, the number of passes $k = 2$. A 9 x 1.0mm tube can be drawn from a rolled tube blank of 28 x 1.0mm in 4 passes; for this purpose the TCRM must include two blocks.

COMPARISON OF THE OUTPUTS FROM MULTI-STRAND MILLS AND HIGH-SPEED DRAWING BLOCKS

We know from reference (86) that a technology for drawing copper tubes in which line type blank drawing mills are not used is being employed in the USA. Tubes with a diameter of 51mm, a wall thickness of 2.5mm and a length of about 70m are extruded, in a horizontal press with a maximum extruding force of 4500 tonnes, from 254mm dia. ingots 559mm long (weighing about 250kg). These tubes are then drawn down to their final dimensions using high-speed blocks with a diameter of 2134mm and self-positioning mandrels, and are then straightened, cut into measured lengths, and delivered in coils. The maximum length of these tubes is 2195m.

The tubes are extruded into water, and have "blind" front and tail ends, thus eliminating the necessity for pickling before drawing. Owing solely to the extrusion of these thick walled blanks, it is possible to dispense with line type blank drawing mills, and the great length of the blanks makes the use of blocks an economic proposition. The use of large diameter blocks reduces the extent to which tubes become oval when coiled, or entirely eliminates this. On going over to the new technology it has been possible to replace 33 line type mills by nine drawing blocks with the same output, thus creating a consecutive technological production flow in the shop.

Let us compare the output of blocks of this type with those of a "Spidem" type mill drawing 22 x 1.5mm tubes from 50mm dia. extruded tube blanks with 2.5mm thick walls. A similar comparison could also be made for other chain-operated mechanized drawing mills.

We will take a procedure for drawing tubes in five passes with a mean reduction per pass $\mu = 1.3$. Finished tube length is 10m, and the drawing speed in the "Spidem" mill is 42m/min. Drawing is carried out with a self-adjusting mandrel. Table 30 indicates how the dimensions of tubes change in the passes.

Table 30

Alteration in Tube Dimensions by Passes when 25 x 1.5mm Tubes are Drawn from 50 x 2.5mm Blanks in five passes.

Pass no.	Length m	External diameter mm	Wall Thickness mm
Blank	2.69	50.0	2.5
1	3.50	41.8	2.2
2	4.55	35.6	2.0
3	5.90	30.5	1.8
4	7.70	25.8	1.65
5	10.0	22.0	1.5

According to Table 26, the output of a "Spidem" mill used for drawing by the same procedure (I) in three strands is 606m/h.

In the case of a block and a blank 70m long (tube length after 5 passes 259m) and a drawing rate of 400m/min, taking $t_{vsp} = 1.25$mi[n] t_Σ can be determined using equation (107):

$$t_\Sigma = \frac{\ell_{tr}}{v} \frac{\mu^k - 1}{\mu^{k-1}(\mu - 1)} + kt_{vsp} =$$

$$\frac{259}{400} \frac{1.3^5 - 1}{1.3^4(1.3 - 1)} + 5 \cdot 1.25 = 8.30min$$

In this case equation (105) gives us an output of:

$$M = \frac{60}{t_\Sigma} \ell_{tr} = \frac{60}{8.30} 259 = 1870m/h;$$

this is three times as great, although when the block is used the auxiliary time is three times the equivalent figure for the "Spidem mill. Thus one high-speed block similar to the blocks under consideration can replace three three-strand line type "Spidem" mills.

If in an existing shop, it is only possible to produce extruded blanks with a length of about 25m, from which a finished tube 92.5m is produced in 5 passes, then

$$t_{\Sigma} = \frac{92.5}{400} \frac{1.3^5 - 1}{1.3^4(1.3 - 1)} + 5 \cdot 1.25 = 6.98 min$$

$$M = \frac{60}{6.98} 92.5 = 795 m/h$$

With relatively short tube blanks, therefore, the output of a block of this type is 31.6% higher than that of a "Spidem" mill when drawing is carried out without an intermediate cutting operation. When drawing is carried out in a "Spidem" mill, with the tube cut in half after the third pass (procedure III in Table 26), the output of the chain-driven mill (774m/h) is about equal to that of the block.

In the latter case, it will be more rational to instal a "Spidem" chain-operated mill than a block of the type indicated above, since the latter occupies a larger area.

CALCULATION OF RATIONAL DRAWING SCHEDULES

Forces and permissible reductions in Drawing Tubes

In cold drawing the output from a multi-strand mill is governed in the end by the number of passes and the number of strands in each pass. The latter depends upon the power of the drawing mill.

The drawing force and number of passes are linked with the permissible reductions. The maximum reduction during drawing is governed by the strength of the tube end to which the drawing force is applied. A normal drawing process is provided with a strength margin coefficient of

$$k_{zap} = \frac{\sigma_{s_k} F_{tr}}{Q} = \frac{\sigma_{s_k}}{\sigma_{vol}} \geqslant 1.1 - 1.2 \qquad (112)$$

where:

σ_{s_k} = the yield stress of the metal in the work-hardened tube

F_{tr} = the tube delivery end section area;

Q = the drawing force

σ_{vol} = the stress in the tube equal to $\frac{Q}{F_{tr}}$

The drawing force can be calculated using the theoretical or empirical equations given in references (1, 14, 15, 22) and (71 etc) respectively. The equations given by I L Perlin* in reference (14) are among the theoretical equations which most fully cover all the parameters of the process:

<u>for drawing without a mandrel:</u>

$$\sigma_{vol} = 1.15\ \ell n\ \mu \left[\sigma_{sr} + f \cot \alpha' (\sigma_{sr} - \sigma_{upr})\right] + \sigma_{upr} \qquad (113)$$

<u>for drawing with a fixed mandrel:</u>

$$\sigma_{vol} = 1.1\ \ell n\ \mu \left[\sigma_{sr} + Cf \cot \alpha' (\sigma_{sr} - \sigma_{upr})\right] + \sigma_{upr}; \qquad (114)$$

here:

μ = the drawing coefficient, equal to $\frac{F_{zag}}{F_{tr}}$ for a pass under consideration;

f = the coefficient of friction;

α' = the reduced angle:

$$\tan\alpha' = \frac{(D_{zag} - D_{tr}) \tan\alpha}{(D_{zag} - D_{tr}) + 2\ \ell_p \tan\alpha}$$

here:

α = the angle between the generating line of the die block and the axis of drawing (the half-angle of the die block);

ℓ_p = the length of the shaping band:

$$C = 1 + \frac{d_{tr} \cos\alpha'}{d_{tr} + t_{zag} + t_{tr}}$$

where:

d_{tr} = the internal diameter of the finished tube;

* As presented I L Perlin's equations are in simplified form.

$t_{zag}\ t_{tr}$ = the wall thicknesses of the blank and the tube;

σ_{upr} = the stress at the boundary between the elastic and plastic zones.

E Zibel's equation can be applied to engineering equations which have been sufficiently well tried in practice:

for drawing tubes without mandrel:

$$\sigma_{vol} = \sigma_{sr}\ \ell_n\ \mu \left(1 + \frac{f}{\sin\alpha\ \ \cos\alpha}\right) \tag{115}$$

and V A Kochkin's equation, given in reference (73), can be applied.

for drawing with a fixed mandrel: :

$$\sigma_{vol} = \sigma_{sr}\ \ell_n\ \mu \left(1 + \frac{f}{\sin\alpha\ \cos\alpha} + \frac{f}{\tan\alpha}\right) \tag{116}$$

When the calculated coefficients of friction found by N P Belousov (Table 31) are used, equations (115) and (116) provide satisfactory accuracy (within ± 10%), and require much less time for calculation by comparison with the more complicated equations given by I L Perlin (14), L E Al'shevskii (15) and other authors

As well as I L Perlin's equation, those of E Zibel, V A Kochkin and certain other authors can be written in the form

$$\sigma_{vol} = \sigma_{sr}\ \ell_n\ \mu A, \tag{117}$$

where;

σ_{sr} = the mean tensile strength of the metal before deformation and after it*

A = the coefficient depending on the angle made by the generating line of the die plates to the drawing axis (α), the coefficient of friction (f), the length of the calibrating band in the die (ℓ_p), and the stress at the boundary between the elastic and plastic zones (σ_{upr}).

Table 31

Alloys	Pass numbers			
	1	2	3	4
L62	0.11	0.11	0.11	0.11
L070-1	0.10	0.11	0.12	-
L68	0.09	0.09	0.12	-
Copper	0.12	0.15	0.15	0.16
MNZh5-1	0.12	0.15	0.15	0.16

The critical reduction μ_{kr} at which a tube breaks can be found using equations (112) and (117). For this purpose, we make the following assumptions:

$$1)\quad \sigma_{sr} = \frac{\sigma_{s_n} + \sigma_{s_k}}{2} = \tfrac{1}{2}\,\sigma_{s_k}\left(\frac{k_u + 1}{k_u}\right)$$

here;

σ_{s_n} and σ_{s_k} = the yield stress figures for the metal before and after deformation;

k_u = the coefficient of strengthening of the metal in a particular pass, $k_u = \dfrac{\sigma_{sk}}{\sigma_{sn}}$

* In references (74) and (14) respectively, N A Shaposhnikov and I L Perlin proved that it is possible in practice to identify the resistance to deformation (the natural rigidity of the metal) with the tensile strength σ_B since these quantities do not differ by more than 5%

2) since $\mu > 0$, $\ln \mu$ can be resolved into series (76):

$$\ln \mu = 2\left[\frac{\mu - 1}{\mu + 1} + \frac{(\mu - 1)^3}{3(\mu + 1)^3} + \frac{(\mu - 1)^5}{5(\mu + 1)^5} + \ldots\right].$$

If we ignore the summands in the square brackets including $(\mu \pm 1)$ to the power of three or more we get

$$\ln \mu \approx 2\left(\frac{\mu - 1}{\mu + 1}\right).$$

If the values of $\ln \mu$ and σ_{sr} are substituted in equation (117), we get

$$\sigma_{vol} = \sigma_{s_k} \left(\frac{k_u + 1}{k_u}\right) \left(\frac{\mu - 1}{\mu + 1}\right) A \qquad (118)$$

It follows from equation (112) that

$$\sigma_{s_k} = \sigma_{vol} k_{zap}$$

If this is taken into account and equation (116) solved for μ, we find that $\ln \mu = B$, or, with an error not exceeding 3.2% with $\mu \leqslant 2.0$:

$$\mu = \frac{1 + B}{1 - B}; \qquad (119)$$

here:

$$B = \frac{k_u}{k_u + 1} \quad \frac{1}{A} - \frac{1}{k_{zap}} \qquad (120)$$

For drawing without a mandrel and $\alpha = 12°$ equations (115) and (116) give us

$$A = 1 + 4{,}97f;$$

for the case of drawing with a short, fixed mandrel,

$$A = 1 + 9.66f.$$

When the critical reductions μ_{kr} are calculated, the margin coefficient $k_{zap} = 1.0$; when the maximum permissible reductions are calculated, $k_{zap} = 1.1-1.2$

When the critical and permissible reductions are calculated for most metals and alloys, the following strengthening coefficients k_u can approximately be used:

Pass no.	1	2	3 and over
k_u	1.83	1.08	1.02

Critical and permissible ($k_{zap} = 1.15$) coefficients of reduction were calculated from the calculated coefficients of friction (see Table 31) and the coefficients of strengthening indicated (Table32). The shape and quality of the driven-in head are largely governed by the maximum reductions, and the values of μ_{kr} found using equation (118) therefore differ slightly from the experimental values given in reference (72) (Table 32).

When thin walled and especially thin walled tubes $\frac{t}{p} \leqslant 0.04\text{-}0.05$ are drawn without a mandrel the degree of deformation of the tubes is limited not by the strength of the tube delive section but by its stability, i.e. its capacity for retaining its prescribed shape. As the reduction is increased there may come a moment when the circular shape of the tube becomes unstable (the critical deformation). In this case part of the outer surface of the tube crumples inwards, and a longitudinal crease forms.

As regards tube stability, the critical reduction can be found from the following equation, given in reference (75):

$$\mu_{kr} \leqslant \frac{1}{1 - 2.8 \cdot 10^2 \left(\frac{t}{D}\right)^2} \qquad (12$$

The shape of the driven in head-clamp has a great effect on the loss of stability by a tube in the draw plate. If "jug"-heads are used, this increases the critical deformation. Drawing through two or three draw plates also makes it possible to increase the deformation per pass from 10 to 25%, with $\frac{t}{D} = 0.018\text{-}0.02$ (77).

With $\frac{t}{D} \leqslant 0.04$, the maximum deformation is governed by the conditions for the strength of the delivery section of the tube or the end of the clamp.

When a schedule is worked out, the aim should be the maximum reduction of tube breakages, which not only reduce the yield of sound metal, but also greatly lower mill output. In the case of single strand drawing, ten breakages in 100 tubes (permitted by the norms in a number of shops at present operating) lower output by 10-20%. When tubes are drawn in three strands in mechanized high-speed mills, when a breakage takes place the drawing speed increases by 50% or more (Table 33). The duration of the drawing speed when a breakage takes place (t'_t) depends on many factors: the number of broken tubes, their length, the point at which they broke etc. It should be noted that, with the same number of breakages when tubes are drawn in three and one strand, the losses of output are far greater in the former case. For instance, with ten breakage per 100 tubes the losses of output when tubes with dimensions of 33 x 30 (1.5) x 5250mm are drawn in three strands using 35.8 x 32.0 blanks are 35%, while they are 15.5% whena single strand mill is use

In general, the reduction in output due to breakages can be

Table 32: Maximum (μ_{kr}) and permissible (μ_r) reductions for different metals and alloys when used for tube production. The Angle of Die Plate Taper α = 12°. Lubricant - soap Emulsion (according to N P Belousov (72))

Metal	μ_{kr}				μ_r				μ_{Σ} dop
	Pass nos.								
	1	2	3	4	1	2	3	4	
M3 copper	(1.89)*	(1.54)	(1.52)	(1.51)	1.50 (1.73)**	1.45 (1.45)	1.35 (1.43)	1.25 (1.41)	3.0-3.5
L070-1	2.30 (1.97)	1.70 (1.67)	1.40 (1.63)	1.30 (1.62)	2.10 (1.80)	(1.55) (1.56)	(1.25) (1.53)	1.20 (1.51)	4.0
L62	1.80 (1.92)	1.50 (1.67)	1.35 (1.64)	1.20 (1.64)	1.70 (1.75)	1.40 (1 56)	1.2 (1.50)	- -	3.0
L68	2.20 (2.06)	1.73 (1.77)	1.46 (1.63)	1.35 (1.62)	2.00 (1.85)	1.60 (1.64)	1.35 (1.53)	1.20 (1.51)	4.0-4.5

Note. The figures in parenthses are the critical and permissible reductions calculated using equation (119) for $k_{zap} = 1.0$* and $k_{zap} = 1.15$**

calculated using the equation:

$$\Delta M\ (\%) = \frac{100}{1 + \dfrac{100}{nq}\,\dfrac{t_t}{t'_t - t_t}}, \qquad (122)$$

where:

t_t and t'_t = the respective cycle times without and with a breakage;
q = the number of breakages per 100 tubes;
n = the number of strands.

The conclusion follows from equation (122) that high strength margin coefficients ($k_{zap} = 1.2\text{-}1.25$) should be specified for the production of tubes in high-output plant; this is confirmed by practice of running mechanized drawing mills.

Factors lowering the limiting values of reduction should be taken into account under works conditions, namely:

Table 33: Effects of tube breakages on the cycle time for a three-strand tube drawing mill Tube metal MNZh5-1 drawing speed 42-45m/min.

Procedure no	Dimensions of initial blank mm	Pass no	Dimensions of tubes before drawing Dxd mm	Dimensions of tubes after drawing DxdxLmm	Reduction μ	Total deformat ε%
		3	35.8x32.0	33.0x30.0x5250	1.365	49.7
		4	33.0x30.0	31.4x28.9x6850	1.255	63.8
I	40x35(2.5)					
		5	31.4x28.9	30.0x28.0x8550	1.310	70.6
II	46x40(3.0)	1	46.0x40.0	41.5x36.8x7600	1.413	29.4

* From stop watch data
** Arithmetical mean of over 85-100 measurements for each schedule

1) the quality of blanks produced by extrusion (variations in tube wall thickness, non-uniformity of mechancial properties along tubes, and plus tolerances);

2) variations in the quality of tube lubrication and the conditions for this during drawing;

3) tool quality (surface finishes and tolerances).

In view of this, the permissible (calculated) reductions must be 8-15% lower than the maximum figure.

An additional cause, which in many cases prevents the maximum reductions from being used, is that the drawing force is increased and , owing to this, the number of strands in a pass is reduced. The procedure can be so worked out as to satisfy the condition

$$Q_{st} \leqslant nQ_{vol} \qquad (123)$$

where:

Q_{st} = the maximum drawing force according to the mill certificate;

n = the number of strands

Q_{vol} = the drawing force

easured rawing tress σ_{vol} N/cm² (kg/mm²)	Tube tensile strength after pass σ_k kN/cm² (kg/mm²)	Strength margin factor $k_{zap} = \frac{\sigma_k}{\sigma_{vol}}$	Drawing cycle, three strands* s	
			No breakages** t'_t	With breakages t'_t
34.2	40-8	1.19	30.8	110.0; 103.8; 46.0
34.8	41.7	1.20	25.4	190.0; 64.8; 63.0
				49.9; 51.0
36.3	42.7	1.18	28.5	70.6; 61.5
30.4	34.3	1.13	34.3	147.0

Table 34

Relationship of the output of a Three-Strand 15-tonne Drawing mill to the number of tubes drawn. Finished tube dimensions 30 x 28 (1.0) x 8055mm. MNZh5-1 Alloy.

Procedure no	Pass No	Size D x d mm		Reductions		Number n of strands	Drawing force Q tonnes	Output m/h		
								Unit blank size "m"		
		Blank	Tube	per pass μ	total μ_{Σ}			1	2	3
I	1	40,0×35	37,4×33,0	1,21		3	8,10			
	2	37,4×33,0	35,4×32	1,35		3	13,56			
	3	35,4×32,0	33,0×30	1,21		3	9,42			
	4	33,0×30,0	31,4×29	1,305		3	11,46			
	5	31,4×29,0	30,0×28	1,25	3,23	3	8,04	707	851	—
II	1	40×35	37,1×33	1,305		3	11,82			
	2	37,1×33	34,2×31	1,375		3	14,79			
	3	34,2×31	31,5×29	1,380		3	13,47			
	4	31,5×29	30,0×28	1,305	3,23	3	9,54	873	1015	—
III	1	40×35	35,6×32	1,540		2	12,02			
	2	35,6×32	32,1×29,5	1,515		2	11,68			
	3	32,1×29,5	30×28	1,385	3,23	3	11,64	851	961 (1)	—
IV	1	46×40	42,2×37	1,250		3	13,68			
	2	42,2×37	39,4×35	1,260		3	14,76			
	3	39,4×35	37,2×33,5	1,250		3	15,03			
	4	37,2×33,5	34,6×31,5	1,275		3	14,79			
	5	34,6×31,5	31,5×29	1,350		3	14,73			
	6	31,5×49	30×28	1,305	4,45	3	10,26	625	784 (4)	782
V	1	46×40	42,2×37	1,250		3	13,68			
	2	342,2×377	38,0×34	1,435		1	7,61			
	3	38,0×34	35,0×32	1,430		2	13,84			
	4	35,0×32	31,5×29	1,330		3	13,92			
	5	31,5×29	30,0×28	1,305	4,45	3	10,28	511	654 (3)	647 (2)
VI	1	46×40	40,9×365	1,515		1	8,04			
	2	40,9×365	36,2×33	1,530		1	8,26			
	3	36,2×33	331,1×305	1,340		3	15,00			
	4	31,1×305	30×28	1,430	4,45	3	13,71	480	623 (2)	731 (2)

Note: The particulars of the "Spidem" mill were used in calculating output. In all the passes v_v = 45m/min and ℓ_{otk} = 3.0m The figures in parentheses are the numbers of the passes after which tubes were cut into two (with m = 2) or three (with m = 3) parts.

Research has proved that, when tubes made of metals and alloys not requiring intermediate anneals are drawn, the most rational procedures are those in which the maximum number of strands is drawn through each of the passes. In order to compare two procedures for drawing tubes from the same blank, (μ_Σ = *const.*) but with different numbers of passes ($k > k_2$) it is necessary to determine the time per tube (the drawing cycle). Obviously procedures will be equivalent as to output if

$$\sum_{i=1}^{k_1} \frac{t_i}{m_i n_i} = \sum_{j=1}^{k_2} \frac{t_j}{m_j n_j} \tag{124}$$

We can see from the figures in Table 34, which gives different procedures for producing 30 x 28 x 8055mm tubes of MNZh5-1 alloy from 40 x 35 (2.5) and 46 x 40 (3.0) blanks, that the most productive procedures are I, II and IV, in which the number of strands in each of the passes is maximum ($n = 3$). When tubes are drawn from 46 x 40 (3.0) blanks of unit length ($m = 1$), reducing the number of passes from six (procedure IV) to five (Procedure V) or four (procedure VI) lowers the output by 18.2 and 22.2% respectively. The reason for this is that it is impossible in a few passes to draw in three strands because the drawing force would exceed the permissible force (15 tonnes according to the mill certificates).

The effectivenesses of procedures can approximately be compared in the following manner. If we assume the drawing cycle times $t_{i(j)}$ for all the passes and in the procedures compared to be the same, it is possible to find and compare the sums $\sum_{i=1}^{k_1} \frac{1}{n_1 m_i}$ and $\sum_{j=1}^{k_2} \frac{1}{n_j m_j}$. The procedure to which the lower sum corresponds is the more rational.

Let us consider an example: We wish to asses whether it is rational to reduce the number of passes from 6 to 4 when drawing a 30 x 28mm tube of MNZh5-1 alloy from a 46 x 40mm blank. The billet factor $m = 2$. In a procedure with six passes (procedure IV in Table 34), cutting is carried out after pass 4, while in the procedure with four passes (procedure V) cutting is carried out after pass 2.

1. Let us claculate the drawing force using equation (116).

We can see from the calculation result, that, when procedure IV is used for drawing, drawing can be carried out in three strands in all the passes with procedure VI, in the first two passes drawing can be carried out in one strand, and in three strands in the following passes.

2. We next find the sum $\sum_{i=1}^{k} \frac{1}{n_i m_i}$:

Procedure IV:

$$\frac{1}{3 \cdot 2} + \frac{1}{3 \cdot 2} + \frac{1}{3 \cdot 2} + \frac{1}{3 \cdot 2} + \frac{1}{3} + \frac{1}{3} = \frac{8}{6}$$

Procedure VI:

$$\frac{1}{1 \cdot 2} + \frac{1}{1 \cdot 2} + \frac{1}{3} + \frac{1}{3} = \frac{10}{6}$$

Procedure IV is thus the more rational. Actually we find from Table 34 that, with $m = 2$, the output figure for procedure IV is 784m/h, while for procedure VI it is 623m/h.

When heavily work-hardened metals and alloys, requiring intermediate anneals are drawn, the rationality of the number of passes is linked with the outlay on annealing. When the dimensions of tube blanks are selected, it is necessary to use as a basis the existing assortment at the works, and the power of the presses and drawing equipment, attempting moreover to provide the minimum number of piercing operations with the minimum number of anneals.

When schedules are calculated, wall thickness must be used as the basis. In the case of tubes with walls $\leqslant 1.5mm$ thick, it is advisable to use tube blanks with wall thicknesses of 2-3mm depending on the grade of the alloy. It is possible to determine the number of passes and the number of anneals necessary to produce walls of particular thicknesses in advance (72).

If, before annealing an alloy, n_0 passes are permissible, and if after the first pass the reduction in wall thickness $\frac{\Delta t}{t}$ is equal to δ_1, if it is equal to δ_2 after the second pass, and if it is δ_n after the last, the thickness of wall (intermediate) t_p which can be reached before annealing will be

$$t_n = t_{zag} (1 - \delta_1)(1 - \delta_2) \ldots (1 - \delta_n).$$

We use $Z = (1 - \delta_1)(1 - \delta_2) \ldots (1 - \delta_n)$; then,

$t_n = t_{zag} Z.$

It is recommended that the following values of Z should be used, when calculating, for different alloys (72):

Alloy	*L070-1*	*L68*	*L62*
Z	*0.36*	*0.38*	*0.54*

After the m-th anneal, the wall of a blank may have decreased to

$$t_{tr} = t_{zag} \; z^{m+1} \qquad (125)$$

where:

m = the number of anneals.

Equation (125) can be used for determining the number of anneals necessary to provide a particular wall thickness t_{tr} from a tube blank with a wall thickness t_{zag}:

$$m = \frac{\log \dfrac{t_{tr}}{t_{zag}}}{\log z} - 1 \qquad (126)$$

Knowing that there can be n_0 passes between anneals, we next determine the number of passes, which is equal to $n_\Sigma = n_0 \; (m + 1)$ This enables us to determine the diameter D_{zag} of the intial blank.

Too great a diametral upset in drawing with a mandrel makes the process difficult ot accomplish. The diametral reduction per pass is therefore taken as:

$$a = \frac{\Delta D}{D_n} = \frac{D_n - D_{n+1}}{D_n} = 0.15 \qquad (127)$$

this corresponds to a reduction

$$\mu_d = \frac{D_n - t_n}{D_{n+1} - t_{n+1}} \approx 1.18$$

Hence:

$$D_{zag} = \frac{D_{tr}}{(1 - a)^{n_\Sigma}} = \frac{D_{tr}}{(0.85)^{n_\Sigma}} \qquad (128)$$

If the tube diameter found is less than the minimum permissible extrusion diameter, the nearest permissible dimension should be taken. In this case the number of passes can be increased, and will be

$$n_{\Sigma} = \frac{\log \dfrac{D_{tr}}{D_{zag}}}{\log (1 - a)} \tag{129}$$

The number of anneals can then be determined:

$$m = \frac{n_{\Sigma}}{n_0} \tag{130}$$

On the basis of the total reduction μ_{Σ} and the permissible total deformation $\mu_{\Sigma total}$ before annealing, the minimum number of anneals necessary can finally be determined accurately using the equation.

$$m = \frac{\log \mu_{\Sigma}}{\log \mu_{\Sigma\ total}} - 1 \tag{131}$$

The reductions per pass must be specified, for instance, on the basis of the permissible reductions per pass (see Table 32). Since the diametral reduction in each pass is taken as ~1.18, the wall reduction will be

$$\mu_t = \frac{t_n}{t_{n+1}} = \frac{\mu r}{1.18} \tag{132}$$

and the reduction will then accordingly be

$$\Delta t = t_n - t_{n+1}\left(\frac{\mu_r}{1.18} - 1 \right) \tag{133}$$

After preliminary calculations, the drawing schedule must be adjusted in order to lighten the individual passes and to anneal at the more rational points in the technological procedure.
Example of the Calculation. It is required to produce 16 x 1.0mm tubes of L070-1 alloy.

1. The initial thickness of the tube blank wall is determined, assuming that the process can be conducted with no intermediat anneal:

$$t_{zag} = \frac{t_{tr}}{z} = \frac{1.0}{0.36} = 2.78mm$$

Since the calculated tube blank wall thickness is less than that of tubes normally extruded, we take $t_{zag} = 3.0mm$. The initial diameter of the blanks is taken as $D_{zag} = 32mm$. In this case the total reduction will be

$$\mu_{\Sigma} = \frac{\left(D_{zag} - t_{zag}\right) t_{zag}}{\left(D_{tr} - t_{tr}\right) t_{tr}} = \frac{(32 - 3)\ 3}{(16 - 1)\ 1} = 5.8$$

and the number of anneals necessary will be (see Table 32);

$$m = \frac{\log \mu_{\Sigma}}{\log \mu_{\Sigma\ total}} - 1 = \frac{\log 5.8}{\log 4.0} - 1 = 0.263 \approx 1$$

Consequently another intermediate anneal is necessary.

2. If we start on the basis of the deformation of the tube wall, in order to obtain a wall 1mm thick from a tube blank with a 3.0mm thick wall the necessary number of anneals, with $z = 0.36$, will be

$$m = \frac{\log \frac{t_{tr}}{t_{zag}}}{\log z} - 1 = \frac{\log \frac{1.0}{3.0}}{\log 0.36} - 1 = \frac{\bar{1}.523}{\bar{1}.556} - 1 =$$

$$\frac{-0.477}{-0.444} - 1 = 0.075 \approx 0$$

The annealing cannot be designated from the tube deformation, and consequently the number of passes will be $n_{\Sigma} = 4.0$ (see Table 32).

3. How many passes are necessary to produce a diameter of 16mm from 32mm? We use equation (129) to determine the number of passes:

$$n_{\Sigma} = \frac{\log \frac{16}{32}}{\log (1 - 0.15)} = \frac{\bar{1}.70}{\bar{1}.93} = \frac{-0.3}{-0.07} = 4.3$$

The conclusion can thus be drawn that the schedule must include one anneal, on the basis of the total reduction, and not less than four passes on the basis of the diametral deformation.

4. We use Table 32 to specify the reductions by passes: in pass 1, $\mu = 2.0$, and in pass 2, $\mu = 1.50$. Since according to calculations, there is one anneal in the schedule, we position this anneal at the centre of the schedule after pass 2. In pass 3, therefore: $\mu = 1.7$ and in pass 4 $\mu = 1.25$.

According to the dimensions of mandrels and the norms, we subdivide as follows by passes:

26 x 32mm blank, $F_{zag} = 273mm^2$
Pass 1, 22 x 25.8mm, $F_1 = 142mm^2$; $\mu = 1.9$
Pass 2, 19 x 22.0mm, $F_2 = 96.5mm^2$, $\mu_2 = 1.48$

Anneal.

Pass 3, 16 x 18.2mm, $F_3 = 59mm^2$, $\mu_3 = 1.64$
Pass 4, 14 x 16.0mm, $F_4 = 47.1mm^2$, $\mu_4 = 1.25$

5. The drawing forces in each of the passes are determined using equation (133) and values of the coefficient of friction from Table 31.

Determining the length allowance on tube blanks for cold deformation (rolling or drawing)

It was pointed out earlier that, when cold processes are used, the yield of sound metal is governed, among other factors (wastage on cutting away tack welds, mill adjustment, defects) by having to cut lengths of tube ends off owing to the allowances on finished tubes and tube blanks. The allowance on the finished tubes is specified in State Standards and Technical Specifications, while those on semi-fabricated products (tube blanks produced by extrusion or cold drawing processes) are governed by standard.

On the basis of constancy of volume it is easy to prove that the yield of sound metal after cold rolling or drawing (without allowing for the metal loss at the clamps) is equal to:

$$G = \frac{1 \pm \dfrac{\Delta t_{tr}}{t_{tr}} \pm \dfrac{\Delta L_{tr}}{L_{tr}} \pm \dfrac{\Delta D_{tr} - \Delta t_{tr}}{D_{tr} - t_{tr}}}{1 \pm \dfrac{\Delta t_{zag}}{t_{zag}} \pm \dfrac{\Delta L_{zag}}{L_{zag} m} \pm \dfrac{\Delta D_{zag} - \Delta t_{zag}}{D_{zag} - t_{zag}}} \qquad (134)$$

here: t_{tr}, t_{zag}, Δt_{tr}, and Δt_{zag} = nominal wall thicknesses and allowances respectively for the finished tube and the blank

D_{tr}, D_{zag}, ΔD_{tr}, and ΔD_{zag} = nominal external diameters and allowances respectively for the finished tube and the blank.

L_{tr}, L_{zag}, ΔL_{tr}, and ΔL_{zag} = nominal length and allowances respectively for the finished tube and blank.

m = the blank number factor.

If it is entirely impossible to produce finished tubes of less than the permissible length, the allowance on the blank for cutting after extrusion is determined using equation (134) with $G = 1.0$

$$\frac{\Delta L_{zag}}{L_{zag} m} = \delta \ell = \frac{\Delta t_{tr}}{t_{tr}} \pm \frac{\Delta L_{tr}}{L_{tr}} \pm \frac{\Delta D_{tr} - \Delta t_{tr}}{D_{tr} - t_{tr}} + \frac{\Delta t_{zag}}{t_{zag}} + \frac{\Delta D_{zag} - \Delta t_{zag}}{D_{zag} - t_{zag}} \quad (135)$$

In this case the length L'_{zag} of the blank must be

$$L'_{zag} = \Delta L_{zag} + L_{zag} m$$

or $$L'_{zag} = L_{zag} m (1 + \delta_\ell)$$

where L_{zag} is the nominal length of the blank, calculated with zero allowances;

$$\Delta L_{zag} = \delta_\ell L_{zag} m$$

When equation (135) was deduced, it was assumed that a finished tube with plus allowances must be produced from a blank extruded with all its dimensions (t_{zag} and D_{zag}) minus.

As an example, we can consider the production of $28\begin{pmatrix}+0.0\\-0.45\end{pmatrix}$ x 1 (±0.1)mm tubes $8950\begin{pmatrix}-0.0\\+50\end{pmatrix}$mm long from 38 (±0.5) x 3.5(±0.35)mm blanks. The total cold processing reduction.

$$\mu_\Sigma = \frac{(38 - 3.5)\ 3.5}{(28 - 1)\cdot 1} = 4.47$$

The nominal length of a double length blank ($m = 2$) is equal (not allowing for the length at the clamps) to:

$$L_{zag} = \frac{8950 \cdot 2}{4.47} = 4000mm$$

On the basis of using cold reduction to produce tubes with consistently plus dimensions ($D_{tr} = 28.0mm$; $t_{tr} = 1.1mm$; $L_{tr} = 9000mm$ from extruded (or rolled) blanks with minus dimensions $D_{zag} = 37.5mm$ $t_{zag} = 3.15mm$) we find the length allowance for the blanks from equation (135):

$$\delta_{\ell} = \frac{\Delta L_{zag}}{L_{zag}\, m} = \left(\frac{0.1}{1.0} + \frac{0.0 - 0.1}{28 - 1} + \frac{50}{8950} + \frac{0.35}{3.5} + \frac{0.5 - 0.35}{38 - 3.5} \right) = 0.206$$

whence: $\Delta L_{zag} = 4000\ (1 + 0.20624) \approx 4855mm$

When the blank length found for drawing (or rolling) tubes with minus dimensions ($D_{tr} = 27.55mm$; $t_{tr} = 0.9mm$) from blanks extruded with plus dimensions (38.5; $3.85mm$) is normed, the yield of sound metal from the process (without allowing for wastage at the clamps) will, according to equation (134), be:

$$G = \frac{1 - \frac{0.1}{1.0} - \frac{50}{8950} - \frac{0.0 - 0.1}{28 - 1}}{\frac{0.35}{3.5} + \frac{855}{4000} + \frac{0.5 - 0.35}{38 - 3.5}} \approx 0.67$$

When tubes with the nominal dimensions are drawn (or rolled) from blanks with plus dimensions, the yield of sound metal will be $G = 0.76$. Taking into account the equal likelihood of other combinations of dimensions, the mean coefficient of the yield of sound metal will be $G_{mean} \approx 0.83$.

Since the probability of drawing (or rolling) tubes with plus dimensions from minus blanks is not high (~ 11%), in order to improve the yield of sound metal it is advisable to norm the drawing (or rolling) of tubes; the dimensions of which decrease from the nominal to the minimum and to extrude blanks with dimensions increasing from the nominal to the maximum. The allowance on the blank can then be determined on the basis of removing the ends of finished tubes 2 x 100mm or less. In the case of the example we are considering, $\Delta L_{zag} = \frac{2.100}{4.47} = 45mm$.

$$\frac{\Delta L_{zag}}{L_{zag}} \cdot 100 = 1.12\%$$ as against 22% in the previous case.

If the third and fourth summands in the numerator and denominator in equation (134) are ignored (they are an order of magnitude lower than the first and second summands), the yield of sound metal can be determined:

for drawing (or rolling) minus tubes from plus blanks

$$G = \frac{1 - \frac{\Delta t_{tr}}{t_{tr}}}{1 + \frac{\Delta t_{zag}}{t_{zag}}} = \frac{1 - 0.1}{1 + 0.1} = 0.82 \text{ (0.80 using the acurate equation)}$$

for drawing (or rolling) minus tubes from blanks with the nominal dimensions

$$G = \frac{0.9}{1.1} = 0.90 \ (0.90);$$

for drawing (or rolling) tubes with the nominal dimensions from plus blanks

$$G = \frac{1}{1.1} = 0.91 \ (0.88);$$

for drawing (or rolling) from blanks with the nominal dimensions

$$G = 1.0 \ (1.0)$$

Since the likelihoods of these cases are the same, the mean coefficient of yield of sound metal will be

$$G_{mean} = \frac{0.82 + 0.90 + 0.91 + 1.0}{4} \approx 0.91 \ (0.90)$$

The standardisation of the limits by which the allowance on blanks varies between the nominal and the maximum plus, and the allowances on tubes between the nominal and the maximum minus thus makes it possible in the case under consideration to raise the yield of sound metal by about 10%. This arrangement of procedures with the customers supplied with products having minus allowances entirely satisfies the problems involved in saving metal in the national economy. The slight increase in tool costs due to the more rigid requirements as to their dimensions and to their shorter lives when used for producing tubes made of expensive non-ferrous metals is completely recovered through the amount of metal saved.

Chapter III

THE COLD ROLLING OF TUBES

Tubes are cold reduced in reciprocating (Pilger-type) mills, for which the abbreviation is KhPT mills. By comparison with other methods of cold reduction for metals, the particular property of rolling in KhPT mills is that the degree of deformation is high (reduction μ_Σ = *6-12 or over*), and this eliminates intermediate annealing operations when a number of alloys are rolled with yields of sound metal reaching 95-98%. Together with the absence of additional operations (for instance knocking off tack welds, etc.) this makes the cold rolling of tubes competitive with drawing. The high degree of deformation attained with KhPT mills is due to the high deformation ratio and the alternating stressed state in the metal (8-11).

Modern methods of groove design (8) have made it possible to roll tubes with sufficiently rigid tolerances as to both external diameter and wall thickness. The initial relative fluctuation in wall thickness for the tube blank is greatly reduced during rolling (10,11), and this makes it possible to produce tubes with extremely little eccentricity of external diameter relative to internal diameter, and with high mechanical properties. Cold rolling has the additional advantage that tubes can be produced with their section varying along their length (13) and with different profiles (6, 16, 77).

INCREASING EFFICIENCY IN THE COLD ROLLING OF TUBES

Factors determining the output of a KhPT mill

With a tube blank feed m and a total reduction μ_Σ, a tube $m\mu_\Sigma$ long is produced in each double pass. The quantity $m\mu_\Sigma$ is usually called the linear displacement of the finished tube. With a number n of double passes per minute for a stand, the output of a KhPT mill will be:

$$M = \frac{60 \cdot n \cdot m\mu_\Sigma \cdot k_{mach}}{1000} \quad m/h \qquad (136)$$

where:

$m\mu_\Sigma$ = the linear displacement in mm;

k_{mach} = the coefficient of machine time

Hence, in terms of finished tubes, mill output is

$$G = M \cdot \gamma kg/h \tag{137}$$

where:

γ = the weight of a metre run of tube in kg/m
The machine hours per ton of tubes rolled is determined using the equation:

$$C = \frac{1000}{M \cdot \gamma} = \frac{10^6}{60 \cdot n \cdot m\mu_\Sigma \cdot k_{mach} \cdot \gamma} \tag{138}$$

The higher the coefficient of machine time k_{mach}. the greater the number of double mill stand passes a minute (n), and the greater the linear displacement $m\mu_\Sigma$ per double mill stand pass, the higher the output of a KhPT mill.

Let us consider the factors governing the magnitude of k_{mach}, n, and $m\mu_\Sigma$. The "coefficient of machine time" k_{mach} depends on the type of tubes being produced, also on the design of the feed assembly. With a feed m and a number n of double passes a minute, the tube blank holder (carriage) moves a distance $m \cdot n$ mm in a minute. If the holder withdrawal distance is ℓ_w the basic machine time t_{basic} for rolling between recharging operations will be:

$$t_{basic} = \frac{60 \cdot \ell_w}{m \cdot n} s \tag{139}$$

In the case of end loading mills, the auxiliary machine time t_{aux} spent during recharging will be:

$$t_{aux} = \frac{60 \cdot \ell_w}{1000\ v_h} s \tag{140}$$

here:

v_h = the speed with which the holder is withdrawn during recharging, in m/min

Hence the total machine (cycle duration) for rolling in KhPT mills with end charging is:

$$t_{total} = t_{basic} + t_{aux.} = 60\ell_w \left(\frac{1}{m \cdot n} + \frac{1}{1000 v_h} \right) \tag{141}$$

Table 35: Cyclogram of the running of KhPT mills with side charging of blanks.

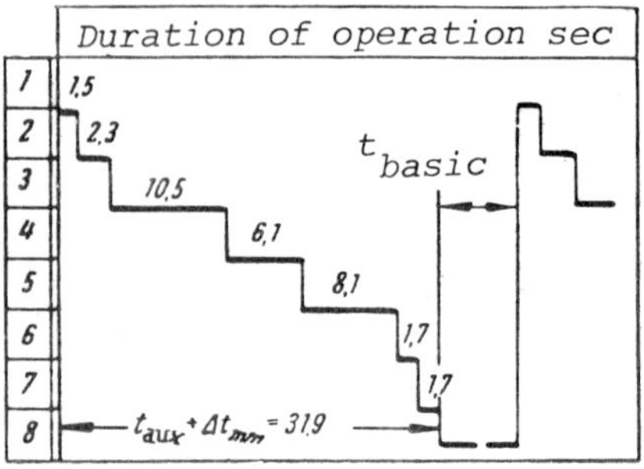

Op.no.	Description of operation
1	Disconnection of feed and mill
2.	Raising chocks
3.	Withdrawing tube blank holder and mill carriage
4.	Charging a blank
5.	Placing holder in working position
6.	Chocking bar carriage
7.	Switching on mill and feed
8.	Rolling

Table 35a: Cyclogram for the running of KhPT mills with end charging of blanks

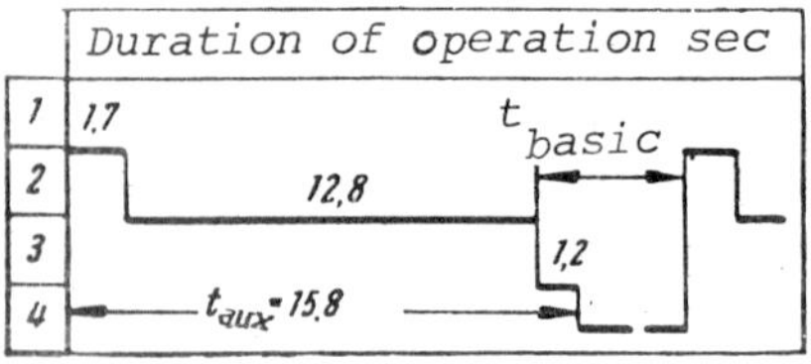

Op.no.	Description of operation
1.	Disconnection of feed
2.	Withdrawal of blank holder
3.	Switching on feed
4.	Rolling

Since $t_{basic} = t_{total} - t_{aux}$, if we divide the left and right hand sides of the equation by t_{total} we get

$$k_{mach.} = 1 - \frac{t_{aux.}}{t_{total}}$$

or, if the values of t_{aux} and t_{total} from equations (140) and (141) are substituted in this equation, we find that the coefficient of machine time for end charging mills is equal to

$$k_{mach.} = 1 - \frac{1}{1 + \frac{1000v_h}{m \cdot n}} \qquad (142)$$

In the case, therefore, of end charging mills, the higher the value of v_h and the lower the values of m and n, the higher the coefficie k_{mach}.

Since the value of $m\mu_\Sigma$ is quite specific (i.e. $m\mu_\Sigma = const$) for the groove design under consideration, from the point of view of raising the value of k_{mach} it is effective to roll with low feeds and high total reductions μ_Σ. Tables 35 and 35a are cyclograms of the running of mills with side and end charging.

In modern end charging KhPT mills (13), the auxiliary machine time $t_{aux} = 0$ and $k = 1.0$. The mills have two feed holders, the operation of which is synchronized: while one holder is feeding a tube into the region of deformation the other is running back into its initial position, and when the first holder reaches the end of its travel the second holder begins to operate, initially synchronized with the first, then independently.

When tubes are produced in side charging KhPT, the total time t_{total} (cycle duration) is made up of the basic machine time t_{basic}, the auxiliary machine time t_{aux} and the auxiliary machine manual time t_{mm}. The time t_{mm} is partially overlapped by t_{aux} and t_{basic}:

$$t_{total} = t_{basic} + t_{aux} + \Delta t_{mm}$$

The tube is deformed during the time t_{basic}. When the mill is being recharged, during the time t_{aux} the bar and blank holder are withdrawn into their extreme position. The value of t_{mm} is governed by the durations of the following basic operations. Charging a blank into the openings in the mill, positioning the holder bar and bringing the carriage up to the end of the blank.

The values of t_{basic} and t_{aux} are found by means of equations (139) and (140), while the continuous part of the time Δt_{mm} is determined with a stopwatch, Table 35 gives values for the cycle time and its components, also the coefficient of machine time for side charging mills.

Although the coefficient k_{mach} is higher for end charging mills, side charging mills have certain advantages in that the surface of the holder can be inspected regularly during each recharging operation and the total length of the plant is less. It is particularly important to know the state of the surface of the holder when metals and alloys which tendto adhere to the tools are rolled (titanium, nickel, cupronickel, etc).

The output of a KhPT mill can be calculated from stopwatch data For instance, in metres of tube blanks rolled, mill output is:

$$M_{zag} = \frac{3600 \cdot \ell_w}{1000\ t_{total}} \quad m/h \qquad (143)$$

hence the output in metres of finished tubes is

$$M_{tr} = M_{zag} \cdot \mu_\Sigma$$

The output of a KhPT mill in weight of rolled stock is

$$G = \frac{3600 \cdot \ell_w}{1000 \cdot t_{total}} \cdot \gamma_{zag} = M_{zag} \cdot \gamma_{zag} \ kg/h$$

In the case of end charging mills the total time t_{total} is directly proportional to ℓ_w (equation 141). The output of these mills does not therefore depend on the length ℓ_{zag} of the blanks or the blank holder withdrawal distance ℓ_w: when the value of t_{total} from equation (141) is substituted in equation (143) the value of ℓ_w decreases.

In the case of side charging mills, t_{total} also depends on the machine manual time not overlapped, Δt_{mm}, which does not depend on the length of the tube blanks. In mills of this design, therefore, as blank length L_{zag} is increased (i.e. as ℓ_w increases) the value of the coefficient k increases, since the proportion of recharging operations in the running of the mill decreases.

The maximum tube blank length L_{zag} is specified in the certificate figures for the mill. The length of the blank is:

$$L_{zag} = u\,\frac{\ell_t}{\mu_\Sigma}$$

where:

u = the number of multiple lengths of finished tube

ℓ_t = the length of a finished tube

When tubes are cut after rolling, the value of $u\ell_t$ must not exceed the length ℓ_{ta} of the receiving table in the mill:

$$u\ell_t \leqslant \ell_{ta}$$

When automatic saws are used for cutting tubes as they travel,

$$\frac{u\ell_t}{\mu_\Sigma} = L_{zag} \leqslant \left[\ell_{cert} \right]$$

where: $[\ell_{cert}]$ is the certified length of the blanks (or the maximum distance by which the blank holder is withdrawn during recharging).

Number of double mill stand passes a minute: The rigidity of a work stand, its design and mass, and the mechanisms for feeding and turning blanks, govern the output of a mill: the feed and total deformation per cycle and the number of cycles (double stand pass) per unit of time. In the KhPT-32, KhPT-55 and KhPT-90 mills the work stands (figs 17 and 19 in reference 13) are in the form of enclosed type housings made of H-sections; this provides the stands with increased rigidity but small mass. A work stand together with the rolls, chocks and the system of driven pinions moves back and forth.

The dimensions of a work stand and its mass are governed by the diameter of the work rolls. As the diameter of tubes to be rolled is increased and the deformation forces also increased (on changing from small to large standard dimension mills), the diameter of the work rolls is increased, and while the design is not changed the mass of the work stand increases in proportion. For instance, in the KhPT-32 mill with 300mm dia. work rolls, the mass of a work stand is 1.9*t*; in the kHPT-75 mill, with 434mm dia. rolls, the stand mass is 7.7*t*, etc. The heavy moving masses cause high dynamic loads in the main drive system, and therefore restrict the number of cycles performed per unit of time. The dynamic loads in the main drive mechanism can be reduced by reducing the moving masses and balancing them.

An extremely important solution, which made it possible to reduce the mass of the moving parts of KhPT mills, was found by Soviet designers. In the KhPT-120P mill the work stand has no moving housing: only the work rolls with the system of bearings, drive pinions and all the structurally necessary components move back and forth. This assembly is moved along special guide rails. The deforming forces from the roll bearings are transmitted to the rails along which run roller bearings, then to the fixed housing. By means of this design the mass of the moving parts in the KhPT-120P mill is as small as 7.3*t*. If the work stand of the KhPT-120P mill was of a design similar to that of the KhPT-75 mill, the mass of the moving parts would be 20-25*t*

Attempts at reducing work stand mass by eliminating the housing should be considered one of the effective solutions to the problem of reducing dynamic loads, particularly in the case of large standard dimension mills. Another effective solution to this problem is that of using the principle of balancing the moving masses.

During a cycle, the main drive bears a highly non-uniform load: the motor (during the first half of the work pass and the first half of the return pass) working under active conditions, twice accelerates the moving masses, and twice (during the second halves of the first and reverse passes) it retards them. Under these conditions the components transmitting movement from the main drive to the work stand (the connecting rod in the connecting rod and crank mechanism and the piston rod in the hydraulic mechanism) are subjected to considerable alternating forces, resulting in their rapidly becoming unserviceable.

The forces in the links of the bearing and connecting rod mechanism can be reduced by locking the dynamic forces in the drive crankshaft with accumulation of energy (flywheel principle) or by other mthods of accumulating the energy of moving masses (fig.56)

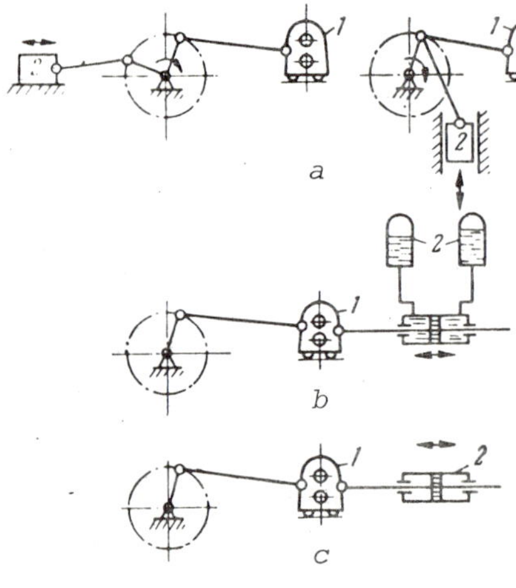

Fig.56: Diagrams of balancing of work stands: 1) work stand; 2) accumulator.

Mannesman-Meyer have used the principle of balancing the moving masses with weights (fig.56a); additional masses (weights) are connected to the drive crankshaft through connecting rods; the bearings of the balancing weights are displaced by 90° relative to the bearings in the stand. When the work stand passes through its extreme positions and its velocity is zero the balance weights equal in mass, are travelling at their maximum velocity and are at the centre of their travel. The energy accumulated by these weights tends to accelerate the work stand from its extreme position to its central position. The stand then enters its period of retardation, and transmits energy for accelerating the balance weights.

The use of balance weights enables Mannesman-Meyer to produce an 1-4½" standard dimension KhPT mill with an increased number of passes:

Mill	KhPT-1½"	KhPT-2½"	KhPT-3½"	KhPT-4½"
Number of double mill stand passes a minute	70-220	60-180	55-165	50-150

The principle of hydraulic accumulation of energy was used by Tube Reduction Co. in their KhPT-18½" mill (fig.56b). The energy accumulator is the hydraulic drive, the pumps in which have the property of reversibility (generator). Spring balancing can also be used, or balancing with a gas accumulator (fig.56c)etc.

The use of the principle of balancing the moving masses complicates mills and increases their dimensions. The requirements regarding the accuracy with which the high-speed assemblies in mills are fabricated and their installation also become more rigid. Balancing however, makes it possible to reduce the dynamic forces with the same number of double mill stand passes, to increase the number of stand passes with the dynamic forces the same, and, very important to instal a main drive with a lower power. As an instance, the power of the drive in the KhPT-18½" mill is 67% less than would be required if there was no balancing.

Promising trends in the development of roll type KhPT mills for rolling tubes made of non-ferrous metals and alloys (principally copper-base and aluminium-base) include using the principle of multi-strand rolling. two, three or more tubes being rolled simultaneously across the width of the work roll barrel. Increasing the number of tubes rolled simultaneously in KhPT mills makes it possible substantially to raise mill output ;the increase in output is slightly less than the increase in the number of strands. This is because the conditions for operating the mill are somewhat complicated (a set of tools has to be made, and they have to be changed and mounted in the rolls- the latter have to be adjusted and the tubes rolled have to be inspected). The particulars of the type KhPT-12-12 three-strand mill designed by VNIIMETMASh (the All-Union Scientific Research Institute for Metal Machinery), produced in the Soviet Union, are given below

Tube blank dia. mm	
maximum	35
minimum	25
Finished tube dia. mm	
maximum	20
minimum	12
Tube blank length mm	
maximum	5
minimum	2
Minimum finished tube wall thickness mm	0.4
Number of double passes a minute in the work stand	
maximum	150
minimum	80
Number of double passes in the work stand a minute at the operating speed	15
Work stand travel length mm	540
Roll diameter, equal to the diameter of the grooves and that of the initial circumference of the driven pinions mm	320
Initial circumferential diameter of the roll driving pinions mm	310
Crankshaft radius mm	270
Connecting rod length mm	1800
Offset mm	120
Power of main electric motor kW	125
Feed per double stand pass (the mill has a gearbox) mm	3;4;6;8; 10.5; 14.5
Maximum travel of feed holder mm	900
Feed holder return speed m/s	0.18
Mill dimensions mm	
length of mill with charging table	32 660
maximum width	3 850
height above floor level	1 391
Weight of mill without electrical equipment kg	51 715

The development of KhPT mills with back-up rolls represents a substantial advance. This makes the mill stands more rigid, while the diameter of the work rolls can be reduced and the permissible pressure exerted by the metal on the rolls increased: the mass of the moving parts in the mill is reduced, and rolling speeds are therefore higher. These mills have excellent prospects for rolling copper-base and aluminium-base alloy tubes.

SELECTION OF THE LINEAR DISPLACEMENT OF THE METAL

The permissible deformation ratio, the permissible pressure exerted by the metal on the rolls, the axial pressure exerted on the blank and the required finished tube quality are decisive in selecting the deformation conditions (reduction and feed).

Permissible deformation: The permissible deformation is determined on the basis of the condition of ductility during the cold rolling of tubes, by the coefficient of deformation ratio n_d. The value of the coefficient n_d is found by using the equation:

$$n_d = \frac{3\ell_{sq}}{m(1 + 2\mu_\Sigma)} \tag{144}$$

here:

ℓ_{sq} = the length of the squeezing part of the groove, in mm

m = the feed in mm

μ_Σ = the total reduction

As the deformation ratio coefficient decreases, the ductility of the metal being processed also decreases, and ultimately it fractures. The limiting value of the deformation ratio coefficient n_d may vary greatly according to the metal being rolled, the deformation conditions and the degree of deformation (Table 36).

The minimum permissible value of the deformation ratio coefficient n_d may vary widely, since these values depend on the flaring of the groove the steepness of the rib, the adjustment of the mill, etc. The data given in Table 36 are characteristic of groove designs extensively used at works.

Equation (144) establishes a connection between the length of the squeezing region in the groove and the speed when tubes made of a particular metal are rolled with a specific degree of deformation.

Measures to ensure increase in ductility and reduce the minimum

Table 36

Limiting (Minimum Permissible) Values of the Deformation Ratio Coefficient n_d for Rolling Tubes in KhPT mills

Groove design	Metal	Deformation	Minimum deformation ratio coefficient n_d	Coefficient n_d from permissible pressure
68x4-41x1.0	Copper	85	6-7	2-2.3
32x3-16x0.7	L62	88	9-10	4-4.5
63x9-38x3.0	LO70-1	78	8-9	4-4.2
60x7.5-38x3	LO70-1	74	7.5-8.5	3.7-4.2

permissible value of the deformation ratio factor n_d were considered in detail in references (16,78). Research has, however, proved that the decisive influence on the ductility of metal during the cold rolling of tubes is exerted by the flare of the groove, or in other words by the ratio of the area of the unsqueezed parts to the area of the squeezed parts of the transverse section of the tube. Table 37 contains schedules used for the experimental rolling of tubes with a view to establishing the deformation conditions with which the metal fails.

When the mechanical properties of the metal (the elongation δ_k) and, the true deformation in an instantaneous region $ln\ \mu_{true} = ln\frac{F}{F}$ are compared, (the section of the groove in which the metal failed in some cases and did not fail in others was investigated), we cannot establish any correlational relationship between δ_k and $ln\mu_{true}$.

When one and the same alloy is rolled under the same deformation conditions with a particular flare, the metal fails (nos.6,8 and 10 in table 37), while as the flare is increased (nos. 5,7,9, in Table 37) the metal does not fail. Analysis proved that this is because, as the groove flare alters, the stressed state in the metal in the groove also alters. The BrA5 alloy (aluminium bronze) was found to be the least ductile of all the alloys investigated (see Table 37). For this alloy it was possible therefore to establish the relationship of the critical deformation in an instantaneous region to the flare of the groove (fig.57). The curve divides the diagram into two parts: region I in which, with any combination of angle of flare and deformation, the metal does not fail in an instantaneous region; region II in which the metal fails with any combination of the angle of flare and degree of deformation in an instantaneous region. The permissible true deformation in the instantaneous region can be established in this manner, also the optimum angles of flare for grooves (this region is shown shaded in (fig.57).

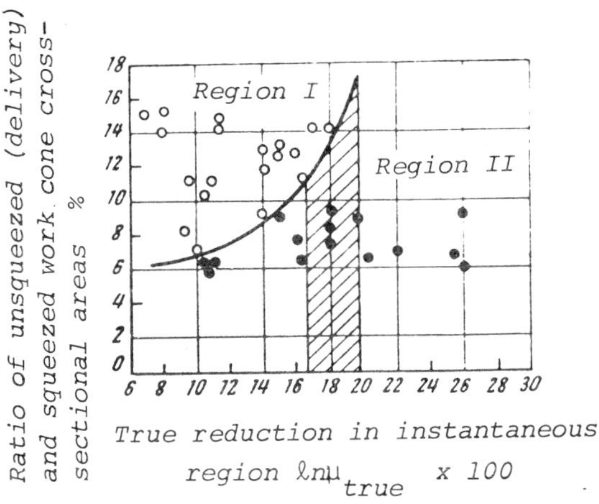

Fig. 57: Relationship of the true deformation in an instantaneous region to the angle of flare of the groove. BrA5 aluminium bronze:
0) the metal did not fail ●) the metal failed.

The principal change in the elongation δ and reduction in area ψ, also in the other mechanical particulars, takes place when the degree of deformation is altered to 50%. As the work-hardening is further increased, these particulars only alter slightly. When tubes are cold rolled, the metal is up to 50% work-hardened at the very start of the squeezing part of the groove; it is therefore true, for practically any groove section, that when the true deformation in an instantaneous region has reached its critical value the metal fails.

Taking the value found for the permissible true deformation with the optimum flare (the shaded region in fig.57) as the mean along the groove, the minimum deformation ratio can be determined using the generally accepted equation

$$n_d = \frac{\ln\mu_\Sigma}{\ln\mu_{true}}$$

The value of the deformation ratio coefficient found using this equation varies, for the optimum angle of flare, over the range n_d = 8-12 (see Table 37), and about corresponds to the values given in Table 36.

When the angle of flare ϕ is about halved, the minimum permissible value of n_d increases by about 50% (see fig.57). When, therefore reference is made to the deformation ratio coefficient it is necessary to bear in mind the effect of the angle of flare on the value of n_d. The rolling schedules given in Table 37 also provide an idea as to the possible deformation and tube cold rolling conditions.

When technological norm charts are compiled, the deformation conditions ($m\mu_\Sigma$) for calculating mill output are determined using the equation

$$(m\mu_\Sigma)_{max} = \frac{1}{2}\left(\frac{31_{sq}}{[n_d]} - m\right) \tag{145}$$

It is recommended that the value of n_d should be taken from Table 36.

The permissible pressure exerted by the metal on the rolls: When tubes made of copper-base or aluminium-base alloys are rolled, also tubes made of other non-ferrous metals and alloys, the output of a KhPT mill is limited by the ductility of the metal and the quality of the tubes rolled. Table 36 gives values of the deformation ratio coefficient n_d calculated on the basis of the permissible pressure. In this case it is 50-67% less than n_d determined on the basis of the metal ductility condition.

The meaned value of the total pressure exerted by the metal on the rolls can be found using the equation given in reference (8) by Yu F Shevakin:

$$P_\Sigma = K_\sigma \sigma_{b50} (D_{zag} + D_t) \sqrt{m\mu_\Sigma (t_{zag} - t_t) \frac{\rho\ mean}{\ell_1}} \tag{146}$$

here:

K_σ = a coefficient depending on the manner in which the metal is strengthened, and has the following value for different metals:

for ferrous metals and alloys	1.42
for non-ferrous metals and alloys	1.0-1.17

$\sigma_{b_{50}}$ = the tensile strength of the metal with 50% deformation (ε) in the groove, kN/cm^2; (kg/mm^2): $\varepsilon = \left(1 - \frac{1}{\mu_\Sigma}\right) \cdot 50\%$

m = the feed in mm;

μ_Σ = the total reduction;

Table 37 Tube Rolling Schedules

Serial nos.	Standard size of KhPT mills	Groove design	Reduction μ_Σ	Metal or alloy	Degree of deformn. ε %		Feed m mm	Parameters of the groove section investigated							Elongation		n_d	Behaviour of the metal **
					Before rolling	After rolling		D_x mm	t_x mm	μ_x	Angle of flare ϕ radians	Δt_x (for a double pass in a stand) mm	Total work hardening $\varepsilon_{\Sigma x}$ %	$\ln \frac{F'}{F} \cdot 100$ %	δ_k	δ*		
1	32-2	38,7×4,3—22,0×0,48	14,3	Aluminium	0	93,0	7,5	23,9	1,0	6,46	0,27	0,60	84,5	56,0	4,0	4,8	4,8	−
2	32-2	38,0×2,8—22,6×0,64	7,0	MZS copper	78	98,3	12,0	24,6	1,39	3,03	0,17	0,67	93,0	44,2	7,0	105	4,4	−
3	32-2	42,1×6,75—22,0×0,75	14,9	MNZh5-1	0	93,3	2,7	23,9	1,37	8,52	0,18	0,30	88,2	20,4	7,5	100	13,2	−
4	32-2	45,6×4,9—20,0×1,59	6,8	MNZhMts 30-0,8-1,0	0	85,3	9,7	22,4	1,67	6,20	0,22	0,36	84,0	26,0	7,5	113	7,3	−
5	32-2	38,0×3,1—23,2×0,93	5,0	BrA5	75	95	7,3	26,8	2,67	1,61	0,23	0,32	82,0	13,8	2,5	6,0	11,7	−
6	32-2	38,0×3,1—23,2×0,93	5,0	BrA5	75	95	7,4	26,2	2,43	1,79	0,19	0,32	82,0	16,0	2,5	5,0	10,0	+
7	55	61,0×5,75—30,0×1,12	9,8	BrA5	0	89,8	4,9	32,5	1,53	7,39	0,26	0,35	86,5	19,5	2,5	4,9	11,7	−
8	55	61,0×5,75—30,0×1,12	9,8	BrA5	0	89,8	4,9	31,2	1,25	9,33	0,17	0,24	89,3	16,5	2,5	5,0	13,8	[illegible]
9	32-2	38,0×3,0—22,2×0,70	7,0	Lo70-1	74,8	96,4	10,0	26,6	2,90	1,54	0,22	0,38	80,0	14,8	3,0	6,5	13,2	−
10	32-2	38,0×3,0—22,2×0,70	7,0	Lo70-1	74,8	96,4	10,0	26,4	2,68	1,65	0,17	0,39	80,0	16,5	3,0	5,4	11,8	+
11	55	58,2×6,8—28,5×1,14	11,2	L68	0	91,1	4,3	29,5	1,38	8,96	0,21	0,29	88,9	20,4	5,0	9,0	11,8	+

*Multiple calculated on the basis of the length ℓ_x of the instantaneous region of deformation

**The plus signs indicate that the metal failed during rolling, the minus signs that the metal did not fail during rolling

$t_{zag}, t_t, D_{zag}, D_t$ = the wall thickness and diameter of the tube blanks and the finished tubes in mm.

ρ_{mean} = the mean radius of the groove rib

$$\rho_{mean} = \frac{D_b}{2} - 0.25\left(D_{zag} + D_t\right)$$

ℓ_1 = the length of the working part of the groove without the sizing region

Equation (146) can be used for determining $m\mu_\Sigma$ on the basis of the permissible pressure P_Σ exerted by the metal on the rolls or the mandrel (8):

$$(m\mu_\Sigma)_{max} = \frac{\left[\dot{P}_\Sigma\right]^2 \cdot \ell_1}{K_\sigma \sigma^2 b_{50}\left(D_{zag} + D_t\right)^2\left(t_{zag} - t_t\right)\rho_{mean}} \qquad (147)$$

Values of n_d calculated using equations (147) and (144) are given in Table 36. We can see from equation (147) that the quantity $(m\mu_\Sigma)_{max}$ is governed by the mean values of the width of the region of deformation

$\dfrac{D_{zag} + D_t}{2}$ the radius ρ_{mean} of the groove rib, and the taper $\dfrac{t_{zag} - t_t}{\ell_1}$

of the groove.

The value of D_t is specified by the groove design, while D_{zag} depends on the taper of the mandrel $2\tan\alpha$, the total reduction as to wall thickness $\mu_0 = \dfrac{t_0}{t_t}$ in the squeezing zone, the technical gap $\Delta_{t\,z}$ between the inner surface of the tube and the mandrel bar, also the reduction ΔD_{red} at the start of the groove.

$$D_{zag} = (D_t - 2t_t) + 2\tan\alpha\ell_1 + t_t\mu_0 + \Delta D_{red}$$

The following conditions must be observed

$$D_{zag} \geqslant D_t - 2t_t) + 2\tan\alpha\ell_{w.m} + \Delta_{t.z} + 2t_{zag} \qquad (148)$$

Here:

$\ell_{w.m}$ = the working part of the groove from the grip to the mandrel cylinder

It has now been established (8,16) that rolling on mandrels with a small taper ($2\tan\alpha=0.002\text{-}0.005$) makes it possible to reduce:

1) the length of the pre-finishing and sizing regions
2) any waviness on the inner surface of the tube such as are inherent in the periodic rolling process
3) the flare of the groove at the end of the squeezing zone and in the pre-finishing zone (i.e. tubes with extremely low ovality can be rolled)
4) mill stoppages due to the breakages of mandrels

The reason for the latter is that, when rolling is carried out using mandrels with a high taper of $2 \tan \alpha > 0.01$, the cylindrical part of the mandrel $d_{ts} = 2 \tan \alpha \cdot \ell_{w.m.} + d_t$ is usually greater than the diameters of the groove at the end of the squeezing zone and in the sizing zone. When these mandrels break or fracture during rolling without protecting rings, considerable forces develop and the bearings become unserviceable, the pinions in the main drive reduction gearing fracture, etc., while the grooves also are damaged. When tubes are rolled with protecting rings, if the mandrel breaks or fractures the protecting ring in the screwdown mechanism has to be replaced, the mill readjusted, and the grooves dressed or the rolls changed. All this causes stoppages of the equipment, sometimes amounting to 10-15% of the total working time.

When tubes are rolled using mandrels with a small taper, the diameter of the mandrel cylinder is 0.8-2.0mm greater than at its squeezing end, and this is less than the diameter of the finished tube. When a mandrel fractures, therefore, no great overload is thrown on the mill. The use of mandrels with a taper $2\tan \alpha = 0.002-0.005$ can, however, only be recommended for end charging mills. In the case of side charging mills, the minimum mandrel taper $2\tan \alpha = 0.01-0.015$, since if $2\tan \alpha$ is lower than this it is difficult to fit the mandrel rod in the working cone during readjustment

The reduction ΔD_{red} of a blank at the start of a groove is governed by the technological gap in the KhPT mill-equation (148), the endurance of the piercing mandrel during extrusion, and the power parameters of the rolling process. From the point of view of better tool (piercing mandrel) life, it is effective to produce tubes with the largest possible aperture diameter. In addition to this, increasing the internal diameter of the tube blank makes it possible to use large diameter bars for holding the mandrel. Research has proved that the axial forces are distributed between the bar and the tube blanks. As the diameter of the bar is increased the proportion of the axial forces applying to the blank decreases, also of course the possibility of the latter welding.

Reductions as to diameter, however, cause additional work-hardening of the metal, and also thicken the tube blank wall thickness by an amount Δt_{red} equal, according to reference (8), to

$$\Delta t_{red} = (0.7 - 0.8)\Delta D_{red} \frac{t_{zag}}{D_{zag}} \tag{149}$$

At the same time, as ΔD_{red} increases, so does the length ℓ_{red} of the reducing zone, the taper of which must not exceed $2\tan\gamma_{red} \leqslant 0.20-0.25$

$$\ell_{red} \geqslant \frac{D_{red}}{2\tan\gamma_{red} - 2\tan\alpha} \tag{150}$$

When non-ferrous metal and alloy tubes are rolled, except for those with low ductilities, the reduction must be between 4 and 15mm. In many cases, however, high reductions are not advisable from the point of view of the ductility of the metal and of reducing the length of the working region of the groove.

Permissible tube blank wall thickness: When the thickness of a tube blank wall is being determined, allowance must be made for the permissible degree of deformation, the resistance of the blank to axial forces, and the possibility of producing good quality tubes. The optimum relationship between the processes of pressing and cold rolling must be borne in mind

During rolling, the ends of the blank do not come together over their entire area, and for this reason the specific axial force q increases and abutting may take place. Increasing the tube blank wall thickness t_{zag} results, on the one hand, in increase in the axial forces, which are a function of P_{Σ} (8.80) and on the other hand in increase in the contact surface area at the butt joint. Reference (8) shows, in fig.150, how P_{Σ}, Q and q vary with relation to the tube blank wall thickness t_{zag} and the gap $\Delta_{t.z}$. Increasing the tube blank wall thickness by 50% makes it possible to reduce the specific axial force in the same proportion, and thereby to avoid abutting.

Research carried out over many years into the cold rolling of non-ferrous metals enables us to recommend the following dimensions for the initial blank (wall thickness): For KhPT-32 mills for rolling tubes with 1.0-1.5mm thick walls the tube blank wall thickness should be 3.0-4.0mm; for the KhPT-55 and KhPT-75 mills for rolling tubes with walls 1.0-1.5mm thick the wall thickness of the blank should be 3.5-5mm, and for rolling tubes with walls 3.0mm thick the wall thickness of the blank should be 7.0-9.0mm

In many cases however, the thickness of the tube blank wall can be much greater than the recommended thickness (e.g. see Table 37)

It must be noted, however, that increasing the tube blank wall thickness may cause the quality of the tubes rolled to deteriorate (increased ovality and faceting). This is due to changes in the relationship between the longitudinal and transverse deformations in the instantaneous region of deformation.

As in the case of rolling plates and section (81, 82) when tubes are cold rolled the principal direction of flow by the metal is governed by the relationship between the deformed and undeformed (or less deformed) zones and by the relationship ψ between the length ℓ_x and the width B_x of the instantaneous region of deformation:

$$\psi = \frac{\ell_x}{B_x} = \frac{\sqrt{2\rho_x \Delta t_x}}{B_x} \qquad (151)$$

Here:

Δt_x = the reduction in the instantaneous region of deformation and is equal to $\Delta t_x = m\mu_x (\tan\gamma_x - \tan\alpha)$;

ρ_x = the calibre radius along the ridge of the groove:

$$\rho_x = \rho_0 - \frac{D_x}{2} \qquad (a)$$

where:

ρ_0 = the radius of the calibre barrel (the roll), $\rho_0 = 0.5D_b$

D_x = the diameter of the groove in section x

At the end of the squeezing region we have:

$$D_x \approx D_t \qquad (b)$$

and

$$m\mu_x \approx m\mu_\Sigma \qquad (c)$$

we will assume that:

$$\tan\gamma_x - \tan\alpha \approx \frac{t_{zag} - t_t}{t_{total}} \xi \qquad (d)$$

and that the width of the contact surface in the instantaneous region of the deformation at the end of the squeezing zone is:

$$B_x = D_t \cos\phi; \qquad (e)$$

here: the presetting angle $\phi = 26°$

Then when the values of (a-e) have been substituted in equation (151) and the latter solved for t_{zag} we find that:

$$t_{zag} \leqslant \frac{\psi D_t^2 \cos^2 \phi \, \ell_{total}}{\xi 2m\mu_\Sigma \left(\rho_0 - \frac{D_t}{2} \right)} + t_t \qquad (152)$$

It was found, on the basis of experimental data, that with a presetting angle $\phi = 26°$, $\frac{\psi}{\xi} \cdot \cos^2 \phi = 0.74$, therefore

$$(t_{zag})_{max} \leqslant \frac{0.74 \, D_t^2 \ell_{total}}{2m\mu_\Sigma \left(\rho_0 - \frac{D_t}{2} \right)} + t_t \qquad (153)$$

Taking the prescribed mill output $m\mu_\Sigma$ and the geometrical tube dimensions D_t and t_t as the basis, equation (153) can be used for calculating the maximum permissible tube blank thickness which can be rolled in a mill of the standard type under consideration to produce tubes with the required quality.

OPTIMUM RANGE OF MILLS

Since the cold rolling of tubes was first used industrially in 1932-1936, this method has come into extensive use for producing small diameter tubes. A large number of KhPT mills have been designed. As the KhPT process has been mastered, the range of tubes rolled has increased as to both dimensions and materials. The range of mills is now extremely wide. At the same time, for a long time (till 1950-1955) the principal working part of the mill (the work stand) remained unchanged nor did the mill speed particulars alter. Since the KhPT method was mastered the standard mill dimensions were not altered (the KhPT-75, KhPT-55 and KhPT-32) There was thus a discrepancy between the range of mills and their particulars, while the extension of the range necessitates considerable differentiation as to dimensions and other features of design. The final result of all this has been that small dimension mills are not rationally used. In the case of most of the range, the mills are not sufficiently used.

The range of tubes rolled has now been defined. The principal (standard) rolling schedules and deformation regimes have been established in paractice. It is now possible to decide upon the optimum range and to plan the principal constructional changes in existing KhPT mills to raise their output and make fuller use of them in accordance with their technical particulars.

When the optimum range with which the fullest use is made of the strength particulars of a mill (the pressure of the metal on the rolls) and the ductility of the metal (fractional coefficient) was established, the following procedure was used (82). Graphs for the relationship of the deformation ratio coefficient n_d to the linear displacement $m\mu_\Sigma$ were plotted, using equation (144), for each standard mill size. The relationship was determined for the maximum and mimimum characteristic values of feed for each type of mill. As a result an extremely narrow range of variation in the coefficient n_d was obtained (see Table 36).

The maximum and minimum values of the sum of the diameters $D_{zag} + D_t$ also of the reduction $\Delta t_\Sigma = t_{zag} - t_t$ were then established for each particular standard size. Curves for $P_{mean} = f(m\mu_\Sigma)$ were then plotted for the possible combinations of $D_{zag} + D_t$ and Δt_Σ using equation (147). As a result of the combination of $n_d = \phi(m\mu_\Sigma)$ and $P_{mean} = f(m\mu_\Sigma)$ the range of tubes (rolling schedule) for which mill output is limited by the pressure or the deformation ratio is determined (82).

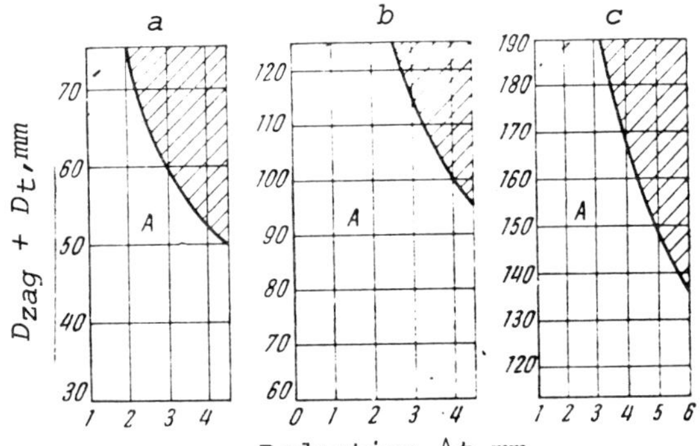

Fig.58: Diagrams defining the regions (shaded) of optimum ranges of L68 alloy tubes rolled (rolling schedules) for different mills:
a) KhPT-32 ($D_{zag} \leqslant$ 42mm $D_t \geqslant$ 16-18mm); b) KhPT-55 ($D_{zag} \leqslant$ 70-62mm $D_t \geqslant$ 21mm); c) KhPT-75 ($D_{zag} \leqslant$ 102mm, $D_t \geqslant$ 48mm)

In accordance with what has been said above, diagrams are given in fig.58a which define the optimum range of tubes for the KhPT-32 mill. The range of tubes (rolling schedules) corresponding to region A should be rolled in mills of other sizes.

The optimum range for KhPT-55 mills: is shown in the diagram in fig.58b. When the range lying in region A is rolled, the mill is under-used by 40-70%.

There are thus boundary schedules the rolling of tubes by which exceeds the technical possibilities (strength) of mills of particular standard sizes, but in this case the strength of another mill will be entirely unsatisfactorily utilized. According to the analysis this covers a considerable proportion of the range. Region A in fig.58b is the region of irrational range (schedules), and tubes must be rolled in mills of other sizes.

Optimum range for the KhPT-75 mill: Figure 58c contains a diagram defining the optimum range (schedules) for the KhPT-75 mill. The schedules lying in region A correspond to mills of other dimensions. Regions of irrational ranges were established by analysis for each size of mill (fig.58). In region A the strength of the mill is only 50% or less utilized. It is important that these regions are not covered by existing standard mill sizes (fig.59)

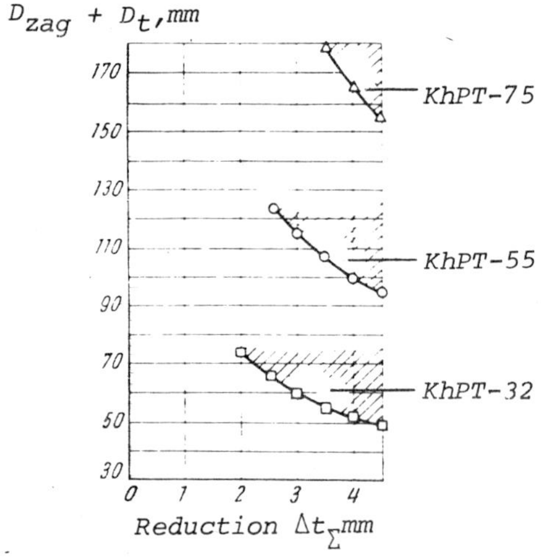

Fig.59: Diagram of the optimum range of L68 alloy tubes for different mills

It is therefore obviously advisable to design KhPT-32, KhPT-55 and KhPT-75 mills of intermediate sizes with lower strength, but higher rolling speeds. This would greatly raise the output and efficiency of the cold rolling of tubes. For this reason it is advisable that works should modernize existing KhPT mills in accordance with the range of tubes and rolling schedules. The latter determines the number of mills requiring modernizing and their dimensions. This would eliminate the lack of co-ordination between the technical particulars of mills and the range of products.

The range of non-ferrous metal and alloy tubes does not at all correspond to the technical possibilities of the existing mills, and this has greatly reduced the efficiency of cold rolling. It is obviously necessary to develop special mills for these metals and alloys.

MILL MODERNIZATION

The following are the most advisable particulars for mills of the intermediate sizes: KhPT-32 with maximum pressure of 0.29-0.34MN (30-35 t), KhPT-55, 0.49-0.59MN (50-60 t), KhPT-75, 1.2-1.3Mn (120-130 t). Let us consider whether it is possible to modernize existing mills (82).

Existing mills must be modernized with the minimum alterations of the components and assemblies in them. This requirement is satisfied by modernization in which the work stand and its assemblies are modified structurally, while the other mechanisms are not altered. Two variations of modernisation can be proposed with regard to the amount and degree of complication or reconstruction, also to its effectiveness.

The first variation involves the minimum amount of alterations: the work rolls together with the radial bearings, the cassettes, the sets of pinions and other components are not changed the work stand housing being cast in a light alloy. The second variation requires a considerable amount of work. The diameter of the work rolls is smaller, and the dimensions of the cassettes, pinions etc. are altered accordingly; the work stand (the housing) is made of cast light alloy, while depending on the method of fixing the grooves the design of the work stand housing may vary.

The fact that the travel through the stand remains constant ($L_k = const$) is common to both methods of modernizing KhPT mills By comparison with the first, the second variation is distinctively more effective, since it provides a greater reduction in the mass of moving parts in the mill. Extensive research has been carried out, followed by production tests, and the variation of modernizing KhPT mills, which is now recognized as being the most rational, has been worked out.

A new design of work roll has been adopted* Instead of a semi-

*(*Yu F Shevakin and F S Seidaliev. Priority Certificate no. 183698, 1966*)

disc, the grooves take the form of a horseshoe (fig.60). This reduces the diameter of the groove by 23-33%. The working length of the groove is slightly increased but the weight of the component is halved. The use of grooves with the new shape has radically altered the design of the work roll (figs.61 and 62). Roll weight is reduced by more than 67%, the amount of labour required to produce the rolls by more than 80%. It is safe to anticipate that the grooves and rolls will have longer lives than the present design.

Fig.60: Groove design for KhPT mills
a) old design; b) new design

Fig.61: Work roll of KhPT mill (general appearance):
a) old design; b) new design

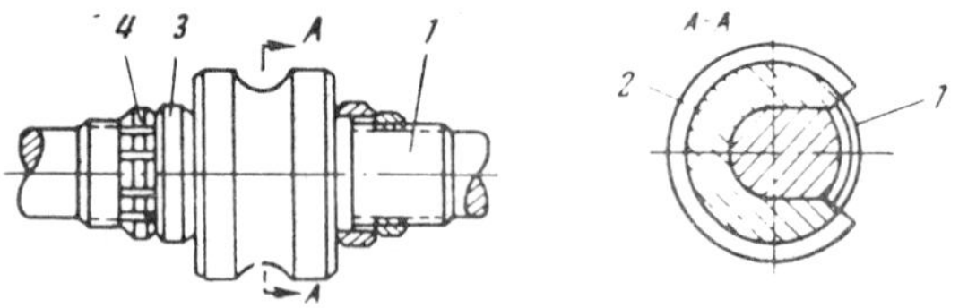

Fig.62: New design of KhPT mill roll system:
1) work roll; 2) horseshoe-shaped groove; 3) ring
4) nut

Table 38

Comparison of the Parameters of Existing and Modernized KhPT mills

Parameters	mill sizes					
	existing			Modernized		
	KhPT-32	KhPT-55	KhPT-75	KhPT-32M	KhPT-55M	KhPT-75M
Work roll (groove) diameters mm	300	364	434	200	280	360
Weight of a set of rolls kg	24	50	94	12	28	45
Weight of work stand kg	2100	4300	7700	500-700	1500-1800	2500-3000
Permissible pressure on rolls MN (t)*	0.75 (76.5)	1.45 (147)	2.07 (210)	0.38/0.65 (40/64)	0.58/0.92 (60/96)	1.2/1.88 (120/192)
Number of double mill stand passes min	80-120	60-90	50-70	<220	<160	<120
Mill stand travel mm	452	625	705	452	625	705
Length of working region of groove mm	400	530	610	400	570	650
Length of roll aperture mm	525	640	700	440	525	640

* The figures to the right of the oblique strokes are the equivalent pressures.

Industrial tests on the equipment showed that, under the same conditions, with the new equipment the pressure exterted by the metal on the rolls by the roll-ing of tubes is 38-44% lower than in the mills at present in use. This has made it possible more than to halve the weight of the work stand.

Table 38 contains the parameters of existing and modernized KhPT mills for comparison. The rolling speed is increased by 100-200%. The modernized KhPT mills can compete with multi-strand KhPT mills.

To modernize existing KhPT mills it is necessary to construct 3-4t of new assemblies. There is no doubt regarding the advisability of the contemplated modernization of KhPT mills at works processing non-ferrous metals and the effectiveness of this is obvious.

The procedure for calculating the output from KhPT mills, when the technological norm charts are compiled, is as follows:

1) the value of n_d is determined from Table 36, or using experimental data;
2) equation (145) is used for finding the linear displacement $m\mu_\Sigma$;
3) knowing the value of $m\mu_\Sigma$, $(t_{zag})_{max}$ is determined using equation (153); this is the metal which can be rolled using a mill of the standard dimension concerned;
4) the optimum tube blank dimensions are determined on the basis of the stressing conditions (see Chapter I). The thickness of the tube blank wall must be less than the thickness determined using equation (153);
5) equation (146) is used for determining the pressure, which must not exceed the pemmissible pressure, otherwise $m\mu_\Sigma$ is determined using equation (147);
6) the dimensions of the tube blank and its advisable length are established in the basis of the optimum ratios for presses and KhPT mills;
7) equations (136/138) or (143) are used for calculating KhPT mill output. The speed particulars of the mills are given in Table 38.

The yield of sound metal when tubes are cold rolled depends principally on whether the tube blank length is correctly selected, and is governed solely by the amount of metal wasted when uneven ends are removed as discards and finished tube ends are removed as being excessive. If the tube blank length L_{bl} is selected on the basis of calculations for the nominal tube dimensions with a reduction $\mu_{\Sigma r}$, and the actual tube dimensions are minus (i.e, the true total reduction $\mu_{\Sigma_{true}}$ is higher than the calculated figure:

$$\mu_{\Sigma\ true} > \mu_{\Sigma r}$$

the yield of sound metal is reduced owing to the necessity for discarding incorrect length tubes. This is the explanation why the yield of sound metal, when tubes are rolled in KhPT mills, is still below unity (100%)

The principal types of rolling defects (lap, depressions, etc.) (16) depend on the quality of the tools and the adjustment of the mill (10). A considerable proportion of the rejections, however, consist of inherited types of defect (double skin, lamination, etc.) for which tube blanks are rejected before being rolled.

PART II

RATIONAL TECHNOLOGICAL SCHEDULES FOR TUBE PRODUCTION

Before proceeding to select rational technologies for producing tubes we will appraise different technologies by comparing them. Let us, for the purpose, consider balanced batches of tubes produced by different methods; these will provide a true picture of the consumption of metal, machine-hours and man-hours on the production of a ton of finished tubes, and it will be possible to determine the technological cost of producing them.

The results of producing similar batches of tubes have shown that, in a number of cases, the accepted consumption coefficients used for calculating the technological schedules differ greatly from the true coefficients. This must be taken into account, since if not, if is impossible to make a correct evaluation of technological schedules and select the most rational. At present, tubes of the same dimensions, made of the same alloys, are produced by different technological schedules at works of the same type. This is largely due to the available equipment in particular shops and to the range and amount of tubes produced. If two shops with different production programmes and producing different ranges of products are compared, it may be found that a particular technological schedule is rational for one shop but irrational for another. Later it will be proved that the degree to which a technological schedule is rational is governed by the range of tubes produced, the production programme and the equipment availabl no schedule can therefore be considered without also taking into account the completed volume of production.

This is also not contradicted by the fact that the technical and economic background for the rational limits of use of each technological process has not yet sufficiently developed, either in general or as applied to particular conditions.

CHAPTER IV

TECHNOLOGICAL SCHEDULES FOR TUBE PRODUCTION AND THEIR ANALYSIS

CONDENSER TUBES OF MNZh5-1 ALLOY (30-20 x 1mm)

Condenser tubes made of MNZh5-1 alloy, their dimensions (30-20) x (1.0-1.5mm) are basically produced by two technological schedules: extrusion type drawing and extrusion type cold rolling.

The technology used at works handling non-ferrous metals for producing condenser tubes with dimensions of 30 x 28 x 8055mm consists of the following processes:

1. Extrusion of the billet in a horizontal hydraulic press (1500- or 3000t)
2. Cutting the extruded billets into measured lengths
3. Pickling in a sulphuric acid bath
4. Fixing the grips at the end of the blanks
5. Drawing in an automatic three-strand 15-t tube drawing mill
6. Straightening the tubes
7. Inspection by the Technical Inspection Department (TID)
8 Annealing in a resistance apparatus.
9. Cutting the finished tubes to a measured length of 8055mm

The "extrusion-cold rolling" schedule which is used omits operations 4-6 and includes the straightening of blanks and their cold rolling down to 28 x 26 x 9000mm in a KhPT-32 mill. The dimensions of the tubes compared are sufficiently close to enable the schedules for their production to be compared.

Calculation and analysis of different methods of drawing blanks with different wall thicknesses: The extrusion/drawing schedule is the one most used at Soviet works in the production of the mass-produced range of non-ferrous metals and alloy tubes. In existing shops, fitted with equipment with particular production equipment (hydraulic press and drawing mill forces and speeds) the problem of improving the output from the basic equipment and reducing the total work required can be solved, for the entire production cycle, by accelerating the processes of extrusion and drawing as much as possible, and by using the optimum combination of these processes. It is no less important to select the optimum production schedule and to provide new shops with the correct equipment, on the basis of the most effective combination of the processes of extrusion and drawing.

Methods of accelerating the drawing process were considered earlier in Chapter II. The extrusion of blanks with specified dimensions can be speeded up by increasing the length of the billet up to the maximum length permissible considering the properties of the alloy, the power of the press and the endurance of the press tools, or alternately by increasing the diameter of the billet and using a larger container (Chapter I). These are the basic arguments used in the analysis which follows.

When extrusion and drawing are considered in conjunction, however, fresh aspects of the optimization of production stages appear. A finished tube with the required dimensions can be produced by drawing blanks with different diameters and wall thicknesses in different numbers of passes. In the case of the "Spidem" type three-strand tube drawing mills, the type considered in our example, the number of tubes drawn simultaneously may vary between 1 and 3. Extruded blanks with different cross-sectional dimensions can be produced by using ignots with different diameter and lengths with greater or lesser extruded tube proportion factors, and with different extrusion process economics.

One can understand that it is effective, in the extrusoon process to produce a blank with the largest possible diameter and wall thickness, and to try to reduce the number of drawing passes, i.e. to use blanks with the smallest possible diameter. These are the conditions which make both the processes as economical of work as possible.

The determination of the optimum relationship between these two tendencies, increasing the dimensions of the extruded billet and reducing the number of drawing passes, the criterion of which is the minimum overall work required, enables us to find the most suitable combination of the processes.

Check calculations were made, using the procedure given in Chapter I, of the force required to extrude blanks for 30 x 28 x 8500mm tubes from different containers in 3000 and 1500t presses. The tube blank wall thickness was varied between 2.5 and 4.0mm The dimensions of the billets were governed by the power of the press and the specified proportions of the tube blank dimensions.

Tube blanks with different dimensions can be drawn down to finished tubes in different numbers of passes; different intermediate cutting schedules can also be used. The tube blank dimensions considered are: 40 x 35 (2.5), 45 x 40 (2.5), 55 x 50 (2.5), 46 x 40 (3.0), 61 x 55 (3.0), 47 x 40 (3.5), 57 x 50 (3.5) and 52 x 44 (4.0)mm. Altogether 85 schedules for drawing and intermediate cutting were considered for the eight different tube blank dimensions (Table 39)

The drawing force was calculated, for a short fixed mandrel, for all the schedules in order to determine the possibility of drawing and the number of tubes which could be drawn simultaneously in each of the passes. The drawing force was calculated using V A Kochkin's equation, this being the simplest equation and providing good agreement with experimental data. The drawing forces found by experimenting when the balance batches were produced did not differ from the calculated forces by more than 4.3%. The procedure given in Chapter II (p.105) was used for determining the drawing cycle time (t_{Σ}) in a "Spidem" mill with the following particulars: $v_V = 45m/min$, $\ell_{otk} = 3.0m$, $t_{zag} = 0.05min$ (Table 40).

A drawing and cutting schedule providing the maximum drawing mill output can be determined for each blank dimension (see Table 40)

Blank dimensions mm	Schedule
40 x 35 (2.5)	4 (4 passes, cut in half after 2nd pass)
45 x 40 (2.5)	12 (4 passes, cut in 3 parts after 2nd pass)
55 x 50 (2.5)	14 (7 passes, twice cut in half after the 3rd and 5th passes)
46 x 40 (3.0)	22 (6 passes, cut in half after the 4th pass)
61 x 55 (3.0)	38 (9 passes, twice cut in half, after the 5th and 7th passes)
47 x 40 (3.5)	50 (8 passes, twice cut in half, after the 4th and 6th passes)
57 x 50 (3.5)	68 (10 passes, twice cut in half, after the 6th and 8th passes)
52 x 44 (4.0)	84 (11 passes, twice cut in half, after the 7th and 9th passes)

The calculations which were made confirmed the following:

1. With all tube blank dimensions, output is minimum when single blanks are drawn.
2. In the case of all blanks for which the calculation was made for cutting in half, these schedules proved to be the most productive by comparison with the other variations of intermediate cutting.
3. In these cases, schedules for drawing blanks cut into three provided the second highest output.

Table 39

Variations of Drawing and Intermediate cutting in the Production of 30 x 28 x 8500mm MNZh5-1 Alloy tubes from different blanks.

Blank dimensions D x d(t)mm	No. of passes	Numbers of schedules for different intermediate cutting patterns				
		No cutting	into 2	Twice into 2	into 3	into 4
40×35	3	5	6 (I)	—	—	—
	4	3	4 (II)	—	—	—
	5	1	2 (III)	—	—	—
45×40	4	10	11 (II)	—	12 (II)	—
	5	8	7 (III)	—	9 (II)	—
55×50	6	17	19 (IV)	18 (II & IV)	20 (III)	—
	7	13	15 (V)	14 (III & V)	16 (IV)	—
46×40	4	27	28 (II)	—	29 (II)	—
	5	24	25 (III)	—	26 (II)	—
	6	21	22 (IV)	—	23 (III)	—
61×55	7	30	—	32 (II & IV)	31 (III)	33 (III)
					34 (IV)	
	9	35	39 (VII)	38 (V & VII)	36 (VI)	37 (V)
47×40	6	44	—	45 (II & IV)	46 (III)	47 (II)
	7	40	—	42 (III & V)	41 (IV)	43 (III)
	8	48	—	50 (IV & VI)	49 (V)	51 (IV)
57×50	7	52	—	54 (II & IV)	53 (IV)	56 (III)
				55 (III & V)		
	8	57	—	59 (IV & VI)	58 (V)	60 (IV)
	9	61	—	63 (V & VII)	62 (VI)	64 (V)
	10	65	66 (VIII)	68 (VI & VIII)	67 (VII)	69 (VI)
52×44	7	70	—	72 (III & V)	71 (IV)	73 (III)
	9	74	—	76 (V & VII)	75 (VI)	77 (V)
	10	78	—	80 (VI & VIII)	79 (VII)	81 (VI)
	11	82	—	84 (VII & IX)	83 (VIII)	85 (VII)

Note: *The figures in parentheses are the numbers of the passes after which cutting is carried out.*

Table 40: Calculation of the durations of single tube drawing cycles and comparison of output with different technological schedules for producing 30x28x8055mm MNZh5-1 alloy tubes (calculated length of tube after drawing 8500mm)

Schedule No.	Blank dimensions	Number of passes	Number of strands per pass, Pass Nos. 1	2	3	4	5	6	7	8	9	10	11	Single tube drawing cycle duration t_{Σ} min	Output of finished tubes m/h	Order of schedule by descending output: For each tube blank dimension	Overall	Increase in output %: By comparison with schedule 1	By comparison with schedule using same length blanks
1	40×35	5	3	3	3	3	3							0,7216	707	6	19	—	—
2														0,5987	851	5	8	+20,4	20,4
3		4	3	3	3	3	3							0,5846	873	3	4—5	+23,5	—
4														0,5025	1015	1	1	+43,6	46,3
5		3	2	2	3	3								0,5920	861	4	7	+21,8	—
6														0,5305	961	2	3	+35,9	11,6
7	45×40	5	3	3	3	3	3							0,5872	869	3	6	+22,9	21,0
8														0,7104	718	6	17	+1,55	—
9														0,6007	849	4	9	+20,1	18,3
10		4	2	2	3	3								0,7077	721	5	15	+1,98	—
11														0,5844	873	2	4—5	+23,5	21,1
12														0,5222	977	1	2	+38,1	35,5
13	55×50	7	3	3	3	3	3	3	3					0,9383	544	7	52	—23,1	—
14														0,6586	775	1	12	+9,62	42,5
15														0,7299	699	4	20	—1,13	28,5
16														0,7123	716	3	18	+1,27	31,6
17		6	3	2	2	3	3	3						0,9529	536	8	56—57	—24,2	—
18														0,6852	745	2	13	+5,37	39,0
19														0,7459	684	5	21	—3,25	27,6
20														0,7220	640	6	29	—9,48	19,4

Table 40 *continued*

Schedule No.	Blank dimensions	Number of passes	Number of strands per pass — Pass Nos. 1	2	3	4	5	6	7	8	9	10	11	Single tube drawing cycle duration t_{Σ} min	Output of finished tubes m/h	Order of schedule by descending output — For each tube blank dimension	Overall	Increase in output % — By comparison with schedule 1	By comparison with schedule using same length blanks
21	46×40	6	3	3	3	3	3	3						0,8159	625	6	31	—11,6	—
22														0,6506	784	1	10	+11,0	25,4
23														0,6518	782	2	11	+10,6	25,1
24		5	3	1	2	3	3							0,9975	511	8	61	—27,7	—
25														0,7801	654	4	25	—7,49	28,0
26														0,7887	647	5	26	—8,49	26,6
27		4	1	1	3	3								1,0647	480	9	67	—32,1	—
28														0,8181	623	7	32	—11,9	29,8
29														0,6973	731	3	14	+3,39	52,3
30	61×55	7	3	1	1	2	3	3	3					1,4566	350	10	82	—50,5	—
31														1,0668	478	8	68	—32,4	36,6
32														1,0113	504	7	62	—28,7	44,0
33														0,9897	515	6	60	—27,2	47,2
34														0,9626	530	5	59	—25,0	51,4
35		9	3	3	2	3	3	3	3	3	3			1,2099	422	9	74	—40,3	—
36														0,8398	607	2	35—36	—14,1	43,8
37														0,8524	598	3	38	—15,4	41,6
38														0,7689	663	1	24	—6,22	57,1
39		9	3	3	2	3	3	3	3	3	3			0,8946	570	4	43	—19,4	35,0
40	47×40	7	3	2	2	2	3	3	3					1,1185	456	11	72	—35,5	—
41		7	3	2	2	2	3	3	3					0,8002	637	5	30	—9,9	39,7
42														0,7509	679	2	22	—3,96	48,9
43														0,8547	597	7	39	—15,6	30,9
44		6	2	1	2	2	3	3						1,2450	410	12	75	—42,0	—
45														0,8336	612	6	33	—13,4	49,2
46														0,9048	564	8	44	—20,2	37,6
47														0,9538	535	9	58	—24,3	30,0
48		8	3	3	3	3	3	3	3	3				1,0551	483	10	65	—31,7	—
49														0,7657	666	3	23	—5,8	37,9
50														0,7092	719	1	16	+1,69	48,8

51														0,7923	644	4	28	—8,91	33,3
52	57×50	7	2	1	1	2	2	3	3					1,5729	324	17	83	—54,2	—
53														1,0601	481	12	66	—31,9	48,5
54														1,1042	462	14	71	—34,6	42,6
55														0,9515	536	9	56—57	—24,2	65,4
56														1,0851	470	13	69—70	—33,5	45,0
57		8	2	1	1	2	3	3	3	3				1,6056	318	18	84	—55,0	—
58														1,0367	492	11	64	—30,4	54,7
59														0,9393	543	8	53—55	—23,2	70,8
60														1,0269	497	10	63	—29,7	56,3
61		9	2	2	2	2	3	3	3	3	3			1,3782	370	16	79—80	—47,6	—
62														0,9214	554	5	47—48	—21,7	49,7
63														0,8401	607	2	35—36	—14,1	64,0
64														0,9228	553	6	49	—21,8	49,4
65		10	3	3	3	3	3	3	3	3	3	3		1,2604	405	15	76	—42,7	—
66														0,9254	551	7	51	—22,1	36,1
67														0,8612	592	3	41	—16,2	46,1
68														0,7889	646	1	27	—8,64	59,5
69														0,8723	585	4	42	—17,3	44,7
70	52×44	7	2	1	1	1	2	3	3					1,7520	291	16	85	—58,9	—
71														1,1544	442	11	73	—37,5	51,9
72														1,0866	470	10	69—70	—33,5	62,2
73														1,2698	402	12	77	—43,2	38,1
74		9	2	2	2	2	3	3	3	3	3			1,3777	370	14	79—80	—47,6	—
75														0,9210	554	6	47—48	—21,7	49,7
76														0,8413	606	2	37	—14,3	63,8
77														0,9234	552	7	50	—21,9	49,2
78		10	2	2	2	3	3	3	3	3	3	3		1,4250	358	15	81	—49,4	—
79														0,9395	543	8—9	53—55	—23,2	51,6
80														0,8570	595	3	40	—15,8	66,2
81														0,9399	543	8—9	53—55	—23,2	51,6
82		11	3	3	3	3	3	3	3	3	3	3	3	1,3711	372	13	78	—47,4	—
83														0,9139	558	4	45	—21,1	50,0
84														0,8348	611	1	34	—13,6	64,2
85														0,9175	556	5	46	—21,3	49,4

4. Of the schedules considered, the drawing output in 18 was higher than that with schedule 1, for drawing a single length 40 x 35 (2.5mm) tube blank.
5. Increasing the tube blank wall thickness by 1mm and its diameter by 7mm [47 x 40 (3.5mm) blank], and increasing the number of passes from 5 to 8 (schedule 50), like increasing the diameter of the blank by 15mm and giving it the same wall thickness [55 x 50 (2.5mm) blank] and increasing the number of passes from 5 to 7 (schedule 16), provide a drawing process with almost the same output as schedule 1 if a multiple length blank is used.
6. With the same cutting schedule, drawing in three strands in all the passes provides higher output than in the case of variations in which drawing is carried out in one or two strands in some of the passes, though the latter makes it possible slightly to reduce the number of passes (see schedules 84 and 72, and schedules 22 and 28, in Table 40).
7. Of all the variations of drawing 46 x 40 (3.0mm) tube blanks, the schedule with six passes and the tubes cut in half after the fourth pass provides the highest output.

In order to obtain the optimum cutting up of extruded tubes and to make it possible to extrude from billets of the maximum length, in many cases different measured cut lengths of extruded tube have been used, with subsequent drawing by different schedules (11 +12, 13 + 14, 15 + 16, 22 + 23, 36 + 39, 66 + 68, 83 + 84, see Table 41) Extrusion of tube blanks in 1500-t and 3000-t presses is envisaged for many schedules (11-12, 22, 37, 38, 49, 50, 68, 83, 84, 83 + 84)

The analysis of 88 variations of drawing for the tube production case under consideration showed that 33 provided the highest output. Therefore only the 33 schedules were analysed (Tables 40 and 41). The data given are of further interest because the calculation of all the variations and their analysis are characteristically highly laborious.

Calculation and comparative analysis of standard process charts for the extrusion type drawing tube production schedules: Standard process charts were compiled for each of the 33 schedules compared, which provide high outputs. The bases of the calculations were as follows:
1. The weight of the billets was determined with the mean diametra tolerance.
2. On the basis of mean measurement results the length of the press discard was taken as 35mm for extrusion with an 1500-t press with an 155mm dia. container and 40mm for extrusion with a 3000-t

press and a 205mm dia. container (See Chapter V)
3. In every case the grips were 200mm long
4. When tubes were produced from blanks with walls 3.5 and 4.0mm thick, also from 61 x 55mm blanks with walls 3.0mm thick, provision was made for initially removing the forged grips. In the case of blanks with walls 2.5mm thick, aslo of blanks with dimensions of 46 x 40mm and walls 3.0mm thick, no provision was made for this operation, since the similar batches confimed the prossibility of using them for drawing right down to the final pass. After the intermediate shearing operation, in most cases it was only necessary to pass the grip under the forging hammers or through forging mills.
5. Allowance is only made for the geometrical wastage during extrusion, shearing blanks, removing grips, and shearing to the final dimensions. This method, used for all the process schedules considered, makes it possible to compare their calculated labour requirements with the minimum consumption coefficients from the standard process charts owing to wastage on adjusting the extrusion and drawing processes and various types of rejection.
6. When the time and production norms were being determined for each of the drawing passes, the calculated cycle time was used (Table 40).
7. The duration of the extrusion cycle was taken, for each billet size, from the effective time norms.
8. When the time and production norms were being determined for pickling, straightening, annealing and cutting to the finished dimensions, the established norms were used as the basis.
9. For the intermediate cutting operations and the subsequent forging of the grip, the time and production norms were determined on the bases of the cycle times for each of these operations, measured during the production of balance batches, and of the equivalent norms used in the shop.

Anlysis of the results of determining the machine work consumed on producing tubes by 33 processes schedules (See Table 40) enables us to draw the following conclusions:
1. The higher the factor for producing extruded tubes in either 1000t or 3000t presses, the lower the specific work consumed on extrusion. For the cases considered, the diagram in fig. 63a shows the manner in which the specific amount of work decreases.
2. Since the extrusion cycle time in a 3000t press is about double the equivalent time for a 1500t press, although it is possible to produce much longer extruded tube blanks in the former, even if the extruded tube factor for it is doubled the specific extrusion work remains about the same. It can therefore be asserted that, depending on the amount of work required of the presses in a shop, there are equally good grounds for producing medium range MNZh5-1 alloy tube blanks in either a 1500t or a 3000t press, provided

that the extruded tube factor for the latter is not less than doubled.

3. As the drawn tube factor is increased, the amount of work required for drawing decreases. This relationship is illustrated quantitatively by the diagram in fig.63b. It can, for instance, be noted that the amounts of labour required are the same for drawing tubes with a factor of 2 in six passes, tubes with a factor of 3 in seven, and tubes with a factor of 4 in eight passes.

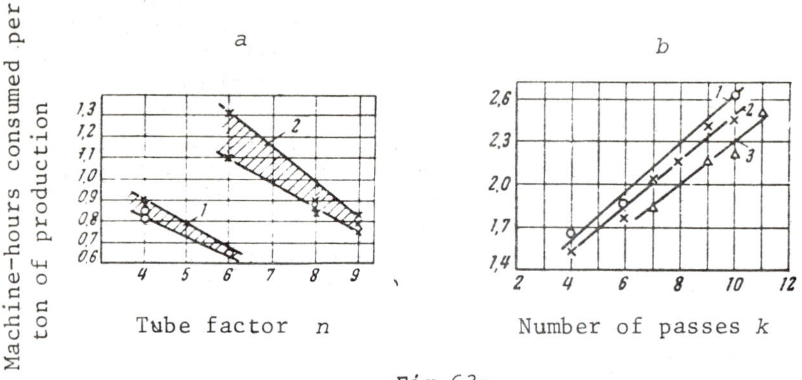

Fig.63:

Man hours consumed on 1t of production with relation to the tube blank factor during the production of 30 x 28 x 8055mm MNZh5-1 alloy tubes:

a) extrusion of blanks in a 1500t press with a container with D_k = 155mm (1); and in a 3000t press with a container with D_k = 205mm (2);

b) the drawing of tubes in a three-strand mill using blanks with a factor m = 2 (curve 1); m = 3 (curve 2); m = 4 (curve 3).

4. The work consumptions on the additional operations (intermediate cutting andfixing the grips) alters substantially (by 2.5--3:1) according to the dimensions (or mass) of the blank and the factor for it, i.e. the number of cuts. This item in the overall labour consumption is extremely important in every case in which a blank with a factor of 4 is used for drawing.

5. The overall work requirements on the process cycles for 21 process schedules (the charts for schedules 1-5 and 13-33 in Tables 40 and 41)and fig.64) are about the same in each case. This means that, when the dimensions of blanks for drawing, presses and numbers of drawing passes are being selected it is only necessary to use as a basis factors such as the load on the equipment and the likelihood of rejections with relation to the dimensions of the blank.

Table 41

Selection of the Billets for extruding blanks for drawing 30 x 28 (1.0) x 8055mm MNZh5-1 Alloy Tubes by the most productive technological schedule (calculated length of tubes after drawing 8500mm)

Schedule No.	Dimensions of blank for drawing mm		Length of extruded tube mm	Billet dimensions mm		Numbers produced from one billet		Length of discard from extruded blank ends mm
	D x d	L_{zag}		D_b	L_b	Blanks for drawing	Finished Tubes	
4	40x35	5500	12 100	150	250	2	4	1000
11	45x40	4900	11 000	150	260	2	4	1200
11+12	45x40	4900+7350	13 000	150	300	1+1	2+3=5	750
11+12	45x40	7350+4900	17 900	200	240	1+2	3+4=7	750
12	45x40	7350	16 100	200	220	2	6	1400
14	55x50	7950	8 950	150	270	1	4	1000
13+14	55x50	7950+2100	10 800	150	320	1+1	4+1 5	750
16	55x50	6000	12 750	200	220	2	6	750
15+16	55x50	6000+4000	14 900	200	250	1+2	3+4=7	900
22	46x40	4100	9 500	150	270	2	4	1300
22	46x40	4100	17 750	200	280	4	8	1350
22+23	46x40	4100+6200	11 100	150	310	1+1	2+3=5	800
36	61x55	4700	15 400	200	330	3	9	1300
36+39	61x55	4700+3150	8 800	150	350	1+1	3+2=5	950
37	61x55	6200	7 100	150	290	1	4	900
37	61x55	6200	13 250	200	290	2	8	850
38	61x55	6100	7 100	150	290	1	4	1000
38	61x55	6100	13 250	200	290	2	8	1050
49	47x40	5300	11 500	150	370	2	6	900
49	47x40	5300	17 000	200	310	3	9	1100
50	47x40	7000	8 050	150	270	1	4	1050
50	47x40	7000	15 050	200	280	2	8	1050
66	57x50	2950	15 500	200	350	5	10	750
66+68	57x50	2950+5700	15 500	200	350	1+2	2+8=10	1150
67	57x50	4300	13 900	200	320	3	9	1000
68	57x50	5700	6 500	150	280	1	4	800
68	57x50	5700	12 450	200	290	2	8	1050
83	52x44	4250	9 500	150	390	2	6	1000
83	52x44	4250	13 900	200	320	3	9	1150
84	52x44	5600	6 550	150	280	1	4	950
84	52x44	5600	12 350	200	290	2	8	1150
83+84	52x44	4250+5600	14 850	200	340	2+1	6+4=10	750
83+84	52x44	4250+5600	16 350	200	370	1+2	3+8=11	900

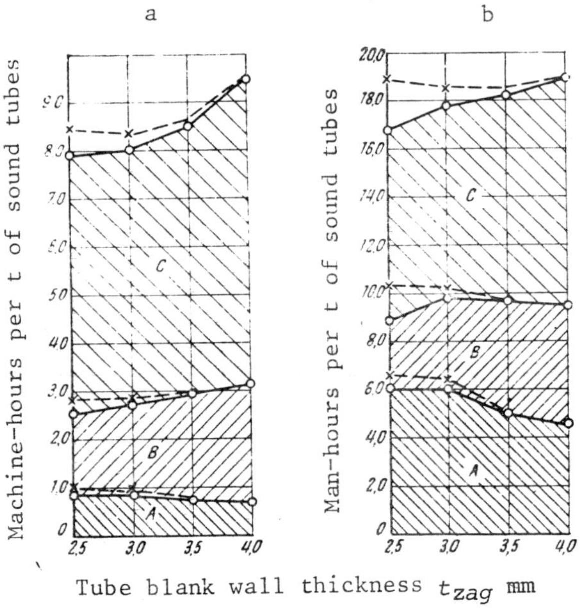

Fig.64: Consumption of (a) machine-hours and (b) man-hours on producing 30 x 28 x 8055mm MNZh5-1 alloy tubes from blanks with different wall thicknesses t_{zag}:
A) extrusion (1500t press); B) drawing (3-strand mill); C) additional consumption (on pickling, removing grips, cutting, annealing etc.)
The continuous lines are for data not including an allowance for rejections, the interupted lines for data allowing for rejections for skin holes and burns through.

The question of the likely amount of rejections is particularly important when the dimensions of the blanks for drawing are being selected. Here we refer only to rejections for double skin and burning through, these being the most usual types of rejection for MNZh5-1 alloy tubes.

It was proved in Chapter I that, during extrusion, two opposite processes take place in the region of deformation: the metal is cooled in contact with the tools and heated by the work deformation. Which of these processes predominates depends on the extrusion speed and the physical properties of the metal and the tool. In the case of the range of MNZh5-1 tubes under consideration on leaving the region of deformation the temperature of the tube has risen by 30-40°C (see Chapter I). An extruded tube wall thickness t_{zag} = *2.5mm* is now the minimum which can technologically be produced.

Owing to the high extrusion stresses, billets are heated to temperatures of 940-950°C. If the heating of metal in the region of deformation is added to this, it becomes obvious that rejections for burns through are very likely. Practice has shown that, as the extrusion temperature is raised (with the other parameters of the process unchanged), the likelihood of rejections for double skin also increases. Increasing the extruded blank wall thickness makes it possible to lower the temperature to which billets are heated, and thus reduce the possible amount of rejections for the defects mentioned above.

As regards the other types of rejections which are encountered - curved, and no-mandrel ("idle") tubes, these are equally possible when tubes are drawn from any type of blank. The number of breakages of grips and tubes during drawing, also the number of incorrect length tubes with an effectively compiled drawing technology, the required sophistication of production and observation of the established tolerances on the dimensions of extruded blanks and tubes during intermediate shearing, must be low and about constant for any size of blank drawn. None of these types of rejections will therefore be taken into account in the rest of this section.

The above arguments made it possible to assume that, when the transverse cross-sectional dimensions of blanks for drawing are increased, the amount of rejections for double skin and burns through in finished tubes must decrease. These assertions have been confirmed by special observations and by the results of producing similar batches of tubes. Data for the amount of rejections (double skin + burns through) are collated, for all these batches, in Table 42. The assertion can be made, on the basis of the results for the 18 batches, that using blanks with greater transverse cross-sectional areas and wall thickness of 3mm reduced the amount of rejections for double skin and burns through from a mean 9.37% to a mean 6.04% per ton of finished tubes.

On the basis of the final result given in Table 42, standard process charts were compiled for three compared systems for producing 30 x 28 x 8005mm tubes: from 40 x 35mm blanks of 1-factor (chart I) and 2-factor (chart II) length blanks, also from 2-factor length 46 x 40mm blanks (chart III). The results of calculating the standard process charts are given in Table 43.

Table 42

Amount of rejections for double skin and burns through when MNZh5-1 Alloy tubes are produced from different types of blank

Batch no	Numbers of tubes: Delivered to TID	Numbers of tubes: Rejected for double skin or burns through	% of rejections for double skin and burns through	Remarks
	40 x 35mm blanks			
1	272	22	8.09	For a 30 x 28mm finished tubes
2	232	42	18.10	
3	122	8	6.55	
4	331	15	4.53	
5	273	12	4.39	
Totals	1230	99	8.05	
6	277	30	10.83	For a 25 x 23mm finished tube
7	371	38	10.23	
8	256	33	12.90	
Totals	904	101	11.16	
Overall Totals	2134	200	9.37	
	46 x 40mm blanks			
9	263	5	1.90	For a 30 x28mm finished tube
10	89	0	0.	
11	120	7	5.83	
12	126	2	1.59	
13	101	9	8.91	
14	133	11	8.27	
15	143	25	17.50	
Totals	975	59	6.05	
16	222	11	4.95	For a 25 x 23mm finished tube
17	121	6	4.96	
18	105	10	9.52	
Totals	448	27	6.02	
Overall Totals	1423	86	6.04	

Table 43

Expenditures on metal and pay per ton of finished 30 x 28 x 8055mm MNZh5-1 Alloy tubes with and without allowance for rejections. (150 x 250mm Billets, calculated length of tube after drawing 8500mm)

Dimensions of blanks	Metal consumption coefficient kg		Cost of metal (Roubles)		Wages (Roubles)		Outlay on metal and wages (Roubles)		Losses due to rejections (Roubles)		
	(a)	(b)	(a)	(b)	(a)	(b)	(a)	(b)	of metal	of pay	total losses
40x35x2850	1575	1440	823-10	812-55	9-15	8-57	832-25	821-12	10-55	0-57	11-02
40x35x5600	1575	1440	823-10	812-55	9-52	8-92	832-62	821-47	10-55	0-59	11-14
46x40x4100	1569	1480	822-63	815-67	9-21	8-82	831-84	824-49	6-96	0-39	7-34

(a) = Allowing for rejections
(b) = Not allowing for rejections.

Table 44 Technical and economic figures for the production of tubes by "extrusion drawing" (30x28x8055mm) and "extrusion-cold rolling" (28x26x7600mm). Schedules (calculated tube lengths: after drawing L_r = 9000m, after cold rolling L_r = 8000m)

Items	Extrusion drawing*			Extrusion cold rolling	
	Chart Nos.				
	I	II	III	IV**	V***
Adjusting coefficient t/t	1.840	1.625	1.622	1.455	1.455
Metal cost roubles/t	843-52	826-75	826-52	813-49	813-49
Pay roubles/t	21-40	21-13	22-03	25-68	15-30
Depreciation roubles/t	8-19	6-81	8-50	47-70	9-61
Outlays roubles/t:					
on electric power	15-49	12-53	16-12	38-70	7-75
on technological tools	0-28	0-25	0-27	3-58	3-67
on running repairs to equipment	1-70	1-53	1-95	14-83	9-84
on maintenance of equipment	0-77	0-63	0-79	4-33	0-88
Total expenditure on production roubles/t	47-83	42-88	49-66	134-82	47-05
Technological cost roubles/t	891-35	869-63	876-18	948-31	860-54
Work $\frac{\text{machine h}}{\text{man h}}$:					
extrusion	$\frac{0.966}{6.762}$	$\frac{0.903}{6.321}$	$\frac{0.950}{6.650}$	$\frac{0.820}{5.740}$	$\frac{0.820}{5.740}$
drawing (or cold rolling)	$\frac{2.187}{4.374}$	$\frac{1.771}{3.542}$	$\frac{2.310}{4.620}$	$\frac{9.600}{9.600}$	$\frac{1.771}{1.771}$
auxiliary operations	$\frac{3.203}{4.952}$	$\frac{3.490}{5.427}$	$\frac{3.484}{5.318}$	$\frac{2.465}{4.175}$	$\frac{2.465}{4.175}$
total for entire technological cycle	$\frac{6.356}{16.088}$	$\frac{6.164}{15.290}$	$\frac{6.744}{16.588}$	$\frac{12.885}{19.515}$	$\frac{5.056}{11.686}$
Output of drawing (or rolling) mill t/h	730	875	765	177	980
Production of finished tubes t/h	0.457	0.565	0.432	0.104	0.565

* Drawing speed 45m/min in all passes

** Single-strand KhPT-32 mill; $m\mu_\Sigma$ = 35.5mm; n = 98 passes/min k_{mach} = 0.87

*** Three-strand KhPT mill; $m\mu_\Sigma$ = 45mm; n = 120 passes/min; k_{mach} = 1.0

Comparison of norm-technological charts I, II, and III shows that the lowest expenditure on metal and pay per ton of finished tubes is provided by the technological schedule for producing 30 x 28 (1.0) x 8055mm tubes from 46 x 40 (3.0) x 4100mm blanks (chart III). The reductions in the expenditures on metal and pay due to the lower rejections with this technological schedule by comparison with the schedules for producing tubes from 40 x 35 x 2850 or 40 x 35 x 5600mm blanks are respectively 0.408 and 0.978 roubles per ton of finished tubes, or 0.49% and 1.17%.

The use, in actual production, of the technological schedule for producing 30 x 28 (1.0) x 8055mm MNZh5-1 alloy tubes from 46 x 40 (3.0) x 4100mm blanks (chart III thus reduces the rejections for double skins and burns through to 6.04%, thereby also lowering the cost of producing tubes.

Calculation and analysis of technological schedules for producing tubes by extrusion and cold rolling: The output of a tube cold rolling mill can be determined using equation (136) for the existing shop schedule for rolling 28 x 26 (1.0) x 9000mm condenser tubes. With $m = 6.3mm$, $\mu_\Sigma = 5.64$ $m\mu_\Sigma = 35.5mm$, $n = 98$ *passes a minute and* $k = 0.847$ (recorded by stop watch), the output of a KhPT-32-2-5 mill (EZTM) will be:

$$M = \frac{m\mu_\Sigma \cdot n \cdot 60 \cdot k}{1000} = \frac{6.3 \cdot 5.64 \cdot 60 \cdot 0.847}{1000} = 177m/h$$

The technical and economic figures for this particular schedule are given in Table 44 (chart IV). The output of the KhPT-32 mill for the schedule under consideration is four or five times lower than for the drawing mill (Chart I, II and III)*.

It was proved in Chapter III that the output of a tube cold rolling mill can first be raised by increasing the linear movement $m\mu_\Sigma$ of the metal, the number n of double mill stand passes a minute (modernization of mills), and the coefficient of re-charging k_{mach}. As table 45 shows, the tube cold rolling mill output achieved at the works we are considering is 15% lower than the maximum output for producing finished tubes.

The coefficient k_{mach} of machine time utilization depends on

---

* *As opposed to the corresponding charts in Table 43, charts I, II and III in Table 44 were calculated on the basis of extruding blanks from billets with dimensions of 150 x 230mm (I) and 150 x 270mm (II and III)*

Table 45

Output of a KhPT-32 Tube cold rolling mill

Works	Metal	Grooving	Reduction μ	Finished tubes		Semi-finished tubes	
				Feed m mm	$m\mu_{\Sigma}$ mm	Feed m mm	$m\mu_{\Sigma}$ mm
A	Copper	34x4÷16x1.0*	8.0	-	-	9.0	72.0
B	L68 brass	42x3.5÷28x1.1*	4.55	-	-	11-12	54.6
		42x3.5÷28x1.0	5.06	8.0	40.5	-	-
C	Copper	34x4÷19x1.0*	6.67	8-9	60.0	-	-
		42x6÷24.2x1.1	8.50	6-7	59.5	-	-
D	Soft aluminium alloy	36x3÷25x(0.8-0.9)	5.11	10.0	51.1	-	-
		36x4÷25x(1.2-1.5)	4.48	14.0	51.0	-	-
		43x3÷32x(0.8-0.9)	4.81	10	48.1	-	-

* End charging mills

the number of tubes produced per blank rolled. Different variations of increasing this latter factor in the cold rolling of tubes were therefore considered. Since the maximum length of a blank is governed by the technical particulars of the mill, the number of tubes per blank can therefore only be increased by increasing the transverse cross-sectional dimensions of the blank. Assuming that the internal diameter of the blank was the same in every case, therefore, the relationship of tube cold rolling mill output to blank wall thickness was obtained at the same time.

Five schedules (a-e) for rolling 22 x 20 (1.0) x 7600mm MNZh5-1 alloy tubes were considered; their technological particulars were as follows:

a) 150 x 285mm billet, 36 x 30 (3.0) x 3100mm blank, factor $u = 2$, feed $m = 8.5$mm, $\mu_\Sigma = 5.26$;
b) 150 x 310mm billet, 38 x 30 (4.0) x 3300mm blank, factor $u = 3$, feed $m = 6.25$mm, $\mu_\Sigma = 7.23$;
c) 150 x 355mm billet, 42 x 30 (4.0) x 3500mm blank, factor $u = 5$, feed $m = 4.0$mm, $\mu_\Sigma = 11.5$;
d) 150 x 375mm billet, 46 x 30 (8.0) x 2500mm blank, factor $u = 5$, feed $m = 2.8$mm, $\mu_\Sigma = 16.2$;
e) 150 x 430mm billet, 46 x 30 (8.0) x 3000mm blank, factor $u = 6$, feed $m = 2.8$mm, $\mu_\Sigma = 16.2$.

In order to obtain the relationship of tube cold rolling mill output to the number of tubes produced from a blank, the linear displacement was taken as being the same for all the schedules and $m\mu_\Sigma = 45$mm, with a number $n = 98$ of double mill stand passes a minute. The schedules were checked by producing experimental batches of tubes.

From the point of view of the extruding force and the possible cooling of the billets, the features of the variations were verified by calculation and experiments. By comparison with 285mm billet (schedule a), calculated for the speed characteristic of a 1500t press the extrusion time for a 430mm billet (schedule e) is practically the same. The initial extruding force for schedule e, calculated by the procedure set forth in Chapter I, is only 1205t. The calculated extruding force for schedule a is 1410t. The technical and economic particulars of the schedules were compared.

When the consumption coefficients were calculated, only the geometrical discards were taken into account. In accordance with industrial practice, the length of the discard after extrusion is taken as being the same for all the schedules compared. For this reason, as the wall thickness is increased the consumption coefficients also increase. Nevertheless, owing to the increase in billet length the specific extruding work required decreases by 9.2-35.5% from schedule a to schedule e, and this indicates additional advantages possessed by schedules with increased blank factors.

When the tube blank factor is increased from 2 to 6, the output of a tube cold rolling mill increases by 17.5% and the work on rolling decreases by 17.9%. The variation is of the same nature for the total technological cycle work (machine hours reduced by 16.9%, man-hours by 20.4%), and the consumption on processing also decreases, by 16.8%. With schedule e, relative to schedule a the reduction in the overall production cost is 2.67%. The relationships of output, expenditure on processing, and cost to wall thickness and billet factor which were obtained are plotted in fig.65. Increase in tube cold rolling mill output involves increasing the coefficient of utilization of machine time (the coefficient of over-loading).

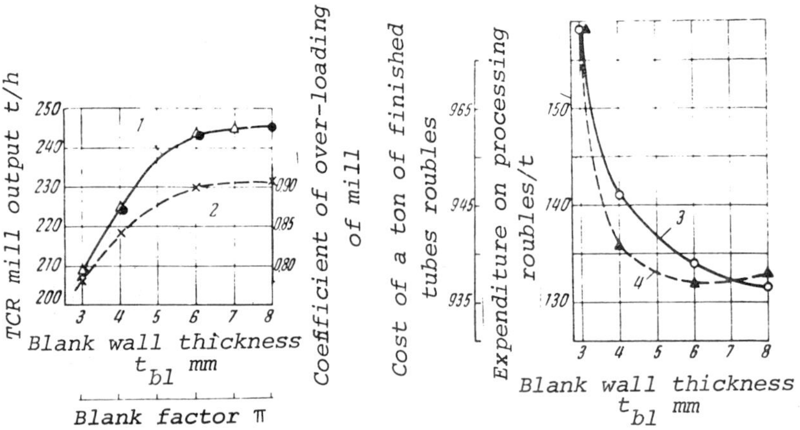

Fig.65: (1) output; (2) over-loading coefficient, (3) expenditure on processing and (4) cost of a ton of finished tubes, for the rolling of 22 x 1mm MNZh5-1 alloy tubes in a KhPT-32 tube rolling mill. Number of double passes n = 98 passes/min. Linear displacement $m\mu_{\Sigma}$ = 45mm

Increase in the tube blank factor involves lengthening the rolling tempo (Chapter III). In our case (for a side changing mill), with small fluctuations in tube blank length the coefficient of machine time increased from 0.794 to 0.919. The re-charging time was taken from stop watch data (T_p = 58s).

Analysis of the results shows that, for the case under consideration of cold rolling from tube blanks with different wall thicknesses, the technical and economic particulars of schedules (output, expenditure on processing, production costs) are greatly improved if the blank factor is increased from 2 to 4-5, corresponding to increasing the wall thickness from 3 to 6mm. When the factor is further increased (the wall thickness increased), the level of these indices is stabilized.

Comparison of the economic effectivenesses of the cold rolling and drawing processes: As an example of the analysis, let us compare all the variations described above as to the production cost of 1 ton of tubes, including the cost of the metal, pay, and running costs associated with running the equipment. The latter, in particular the depreciation costs, the cost of repairs, power consumption and the cost of tools, must be taken into account when technological schedules are compared, since with the "extrusion-cold rolling" schedule these items amount to 80% of the expenditure on processing and 11% of the cost. Comparative data for all the charts are given in Table 44 and fig.66.

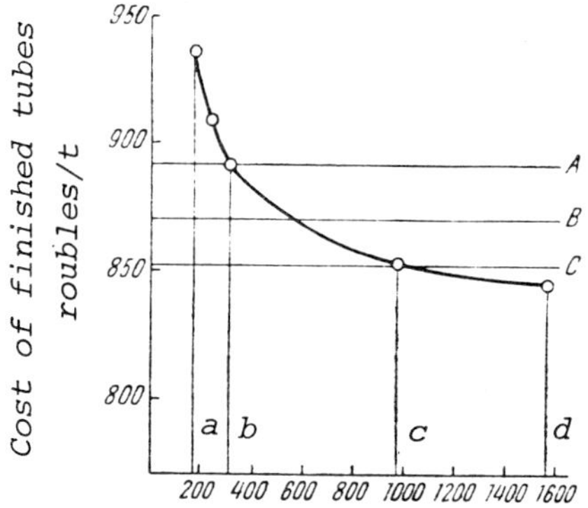

Fig.66: Variation in the cost of finished 30 x 38mm MNZh5-1 alloy tubes to cold rolling mill output: A B C) cost of tubes produced by drawing in a three-strand mill by schedules I, II and III (a, b, c, and d rolling schedules)

It can be seen from the data which have been given that, when the output of a single-strand tube cold rolling mill is increased by 80% the cost of the tubes is 0.3% lower than with drawing schedule I. As published works have more than once pointed out (79), however, high output tube cold rolling mills alone can compete with multi-strand machines. In order, therefore, fully to compare the schedules under consideration, it is necessary to assess the possibilities of high-output tube cold rolling mills (85)

Since no service data have been published in Soviet literature for three-strand tube cold rolling mills, the depreciation figures the expenditures on electric power, and the expenditures on tools (per strand) for these mills were taken as being the same for single-strand mills. In actual fact, the first two particulars are bound to be lower for three-strand mills. We can see from Table 44 that the output of a three-strand tube cold rolling mill, with $n = 120$ *passes/min*, $m\mu_{\Sigma} = 45mm$ *and* $k_{mach} = 1.0$ (see schedule V), is 10-40% higher than that of a "Spidem" drawing machine; costs are 15-64 roubles lower (schedule III) and 30-81 roubles lower (schedule I).

It will, however, be proved later that, by comparison with modern multi-strand "Chevalier" drawing machines, three-strand tube cold rolling mills can only be competitive with 200 passes/min. The maximum possible outputs of single-strand KhPT-32 mills are given in Table 45. The comparison of production schedules for the production of tubes by drawing or cold rolling by the outputs of drawing machines and cold rolling mills cannot, however, fully represent their effectivenesses.

It is necessary also to analyze the technological schedules under consideration in terms of the following indices (calculated per ton of finished tubes):

1) expenditure on metal;
2) the work required for the entire technological cycle and the principal operations, in machine-hours and man-hours;
3) the expenditure on processing and the cost of tubes (according to the components which vary with relation to the technological schedule);
4) the specific capital outlay figures.

The particulars determined on the basis of the results of four-balanced batches with a total weight of 6.67t, are given in Table 46

It follows from Table 46 that, in cold rolling (schedule IV), there are practically no rejections (0.67%), while when drawing is carried out in five passes using 35 x 40mm blanks (schedule I) the rejection are 4.32%. The reason for this lies in the temperature and speed conditions for extruding blanks, which are more favourable to thick wal blanks for tube cold rolling mills, also to breakages of tubes durin drawing. For this reason, also owing to the considerable lengths which have to be cut off finished tubes, in spite of the approximate equal yield of sound metal after extrusion, the overall yield of sound metal is 15% higher after cold rolling than after drawing, and it is 17.7% higher from blanks when rolled than for drawing. This also causes a considerable difference in the adjusting coefficients (Table 44): for cold rolling (schedule IV) the coefficient is 0.405t, or 22% lower than for schedule I. After deducting the cost of the waste metal, this provides a saving, with schedule II, of 31.5 roubles cost of metal per ton of finished tube (see Table 44).

Owing, however, to the comparatively low output of tube cold rolling mills, in spite of the necessity for additional operations when tubes are drawn, the overall work requirements by schedule IV is 81.2% higher in machine-hours and 10.6% higher in man-hours than for schedule I (see Table 44). The reason why tubes produced by schedule IV cost more than tubes produced by schedule I is that depreciation figures are much higher for rolling (by comparison with other works the depreciation figures are greatly exaggerated) than for drawing.

The collated data in Table 44 show the advantages of schedule I: with this schedule the expenditure on processing is 58% less than with schedule IV and allowing for the cost of the metal the saving when schedule I is used, is 46.55 roubles per ton of finished tubes, or 4.9% of their cost. Another important index is the coefficient of capital outlay, which prepresents the capital outlay necessary to provide equal production of each of the schedules. The specific capital outlay when schedule IV is used is 4.9 times the equivalent outlay when schedule I is used.

The data which have been obtained make it possible to conclude that the method, recommended by certain authors, of evaluating the effectiveness of a production schedule solely or principally by the consumed metal coefficient cannot be considered to be sufficiently good. In our case the consumption of metal with the schedule including cold rolling was much lower, but the cost of production was higher than when drawing was used. This is due both to the high cost of the wasted non-ferrous metal (about 90% of the cost of ingots) and to the considerable proportion occupied by the cost of the metal in the cost of tubes (85-95%). For ferrous metals, for which the value of the wasted metal is 50-60% of the cost of the ingots, this situation may substantially alter the picture of comparative effectiveness.

Calculations have proved that, when multi-strand tube cold rolling mills are used, the cold rolling process becomes competitive with drawing in multi-strand machines, even in the case of metals and alloys not requiring intermediate anneals during cold treatment (schedule V in Table 44). In every case

Table 46 Yields of sound metal during the production of balance batches of MNZh5-1 alloy tubes

Schedule	Batch No.	Billet mass t	Extrusion waste		Cutting waste		Weight of extruded tubes t	Rejection*		Weight of finished tubes t	Yields of sound metal %		
			t	%	t	%		t	%		From ingot	During extrusion	From blank
I**	1	1,630	0,261	16,0	0,141	8,65	1,230	0,131	8,05	0,810	49,7	75,5	65,9
	2	1,755	0,307	17,5	0,122	6,95	1,325	0,015	0,85	1,010	57,4	75,6	76,0
	Total for Schedule I	3,385	0,568	16,8	0,263	7,75	2,555	0,146	4,32	1,820	53,8	75,5	71,2
IV***	3	1,755	0,210	11,95	0,160	9,15	1,385	0,022	1,25	1,255	71,5	78,9	90,6
	4	1,530	0,250	16,35	0.130	8,5	1,150	0	0	1,000	65,3	75,2	86,9
	Total for Schedule IV	3,285	0,460	14,0	0,290	8,85	2,535	0,022	0,67	2,255	68,8	77,2	88,9

* The rejections for Schedule I also include breakages of tubes during drawing
** Tube dimensions 30x28x8055mm billet dimensions 150x230mm
*** Tube dimensions 28x26x7600mm billet dimensions 150x250mm

cold rolling has the advantages of considerable mechanization of the operations and the most sophisticated production.

The careful evaluation of the economic effectivenesses of different production schedules, using drawing and cold rolling, which have been made enabled us to plan methods of further improving efficiency. The characteristics of the outlays comprising the cost of production are extremely important when new shops are planned.

SMALL DIAMETER (6-10 x 0.5-10mm) COPPER TUBES

There are various possible high output technological schedules for producing small diameter copper tubes:

1. Extrusion - block drawing
2. Extrusion - drawing with plane draw benches - block drawing
3. Extrusion - cold rolling - drawing

The distinctive feature of the "extrusion - block drawing" schedule is that no line type preparatory draw benches are required. This schedule is used at a works in the USA (86) extrusion being carried out in a 4500t horizontal press; 254mm dia. billets 560mm long (weighing 250kg) are extruded into 51mm dia. tubes with walls 2.5mm thick and a length of about 70m. Extrusion is carried out into water, and it is thus unnecessary to pickle the tube blank before drawing. Since there are plugs at the leading and rear ends of the tubes, their inner surfaces are not oxidized at all, and they can be effectively drawn on self-stabilizing mandrels.

The tubes are drawn on high-speed 2134mm dia. blocks down to their final dimensions, and are then straightened and cut into measured lengths or supplied in coils. Maximum tube length produced is 2195m. The total (overall) reduction is:

$$\mu_{\Sigma} = \frac{2195}{70} = 31.4$$

There are equivalent reductions when 8 x 0.5mm tubes are produced from 70m long 51 x 2.5mm blanks by drawing in nine passes (mean reduction per pass $\mu_{mean} = 1.47$; total reduction $\mu_{\Sigma} = 32.4$)

With a mean drawing rate of 500m/min and an auxiliary time $t_{aux} = 1.0min$ the output of a block in metres of finished tubes can be determined using equation (110). The overall drawing time is:

$$t_{\Sigma} = \frac{\ell_{bl}}{v} \cdot \frac{\mu(\mu^k - 1)}{\mu - 1} + kt_{aux} =$$

$$= \frac{70}{500} \cdot \frac{1.47(1.47^9 - 1)}{1.47-1} + 9 \cdot 1 = 22.7 min;$$

The output is:

$$M = \frac{60}{t_{\Sigma}} \ell_{tube} = \frac{60}{t_{\Sigma}} \ell_{bl} \mu^k = \frac{60}{22.7} \cdot 70 \cdot 1.49^9 = 6000 m/h$$

The great length of the blanks makes it an economic proposition to use blocks, while the use of comparatively thin walled blanks makes line type auxiliary draw benches unnecessary. The use of this schedule for producing copper tubes has enabled one firm in the USA to replace 33 line type draw benches by nine drawing blocks and to create a consecutive technological flow in the shops.

A similar production schedule is now being used at a works in the Soviet Union for producing 8 x 1.5mm tubes from 40 x 3mm blanks. The extrusion is carried out in a 3000t press; the billet dimensions are 245 x 300mm, and each billet weighs 130kg. The billets are heated to 920-950°C in an induction furnace. On leaving the die, tubes drop into a water-filled channel 35m long. To plug the leading end, the length of the mandrel nose after deducting the thickness of the pressure pad must be 5-10mm less than the length of the billet. This means stricter tolerances on billet length when cut.

After back extrusion, the billet is not fully pierced, and at the first moment of extrusion through the die the metal extruded is a rod until the mandrel nose and the die aperture form a circumferential gap. The rear end plug is formed by removing the mandrel nose at the end of the process so that a rod is again extruded throughthe die but the mandrel nose end remains in the metal to prevent air from entering the evacuated chamber within the tube. The output of the press is ~5.5t an hour. After uncoiling, the cooled tubes are transported to the vertical coiling block at the Alma-Ata Works, which has the following technical particulars.

Block diameter mm	1500
Maximum drawing force t	8.6
Drawing speed m/s	0.9 (v_{max} = 540m/min)
Finished tube dia. mm	3-30
Minimum wall thickness mm	0.3
Maximum blank dia. mm	40
Maximum blank wall thickness mm	3

Length of blank m	15-50
Maximum length of finished tube m	2300 (3 x 0.5)

The drawing reductions are 1.47-1.55, and there are practically no cases of tubes breaking. The use of heavy coils (weighing up to 120kg) raises drawing block output. The drawing of these coils, however, requires further mechanization of the blocks in order entirely to eliminate all the manual operations. After drawing, tubes are taken for annealing and finishing operations. The new method of bright annealing copper tubes in a resistance apparatus makes this process extremely economic.

One of the advantages of coil drawing is the reduction in fluctuations in tube wall thickness, since when drawing is carried out on a floating mandrel there are considerable overall diametral reductions. For instance, when 8 x 1.5mm tubes are drawn in six passes from 40 x 3mm blanks, the variation in wall thickness is reduced from ±20% to ±2.5%.

The schedule can be further improved by fitting a longer water channel (up to 50m long), enabling the weight of the billet and coil to be increased to 180kg, press and block outputs to be increased (the former to 25t a shift), and the yield of sound tubes to be raised to 80% or over. The schedule considered has made it possible to dispense with the preparatory line draw bench which supplied coils weighing 60kg (tube length 50m) for the draw block.

In some cases, when short extruded blanks are used (less than 10-12m long) the inclusion of preparatory draw benches in the production line makes it possible to accelerate production and raise drawing block output through increasing the proportion of machine time. This type of technology is rational if the press is the bottleneck; the extrusion of thick walled blanks makes it possible to increase the dimensions (weight) of the billets and to raise the press plant output.

Blanks up to 50m long for blocks can be produced in a three-strand "Chevalier" draw bench with the following technical particulars

Drawing force MN(t)	0.6(60)
Useful length of bench (chain) m	50
Drawing speed with a force of 0.6MN (60t) and a tube blank diameter of 100mm, m/min	8-120
Rate of carriage return m/min	180
Number of Tubes drawn simultaneously	1-3
Maximum tube blank external diameter mm	160
Minimum tube blank external diameter mm	35

Maximum finished tube external dia.	mm	150
Minimum finished tube external dia.	mm	30
Maximum tube blank length	m	40

The longer the tube in each of the passes, the higher the output of line chain draw benches. Obviously when long tubes (more than 15-20m long) are produced, intermediate operations of cutting are not advisable owing to the difficulty of transportation, and to the greater amount of work required for cutting tubes and forging the grips (if the draw bench is not provided with a master, forging, unit). The maximum length of extruded blank transported to the preparatory draw bench is governed by the same factors.

Tubes 50m long can be produced from a 10m long blank in four passes*. The total reduction in this case is $\mu_{\Sigma} = 5.0$, and the mean reduction per pass $\mu_{mean} \approx 1.5$. The output of a "Chevalier" type mill, in metres of finished tube, can be found using equations (101) and (102): the overall time to draw a finished tube is:

$$t_{\Sigma} = k\left[t_{gr} + \ell_{acc}\left(\frac{1}{v'_d} - \frac{1}{v_d}\right) + \frac{\ell_{br} +}{v'_{ret}} + \frac{\ell_{in} - \ell_c}{v_{ret}} + \frac{\ell_{in}}{0.5v_v}\right] + L_t\left(\frac{1}{v_v} + \frac{1}{v_{ret}}\right) \times$$

$$\times \frac{\mu_{mean}^k - 1}{(\mu_{mean} - 1)\mu_{mean}^{k-1}} = 4\left[0.5 + 3\left(\frac{1}{60} + \frac{1}{120}\right) + \frac{3}{40} + \right.$$

$$\left. + \frac{5-3}{180} + \frac{5}{0.5\cdot 120}\right] + 50\left(\frac{1}{120} + \frac{1}{180}\right)\frac{1.5^4 - 1}{(1.5 - 1)\cdot 1.5^3} = 5.14\ min;$$

the output in metres of finished tubes is:

$$M = 3\cdot\frac{60}{5.14}\cdot 50 = 1750m/h$$

* If it is not convenient to transport a long blank through the shop, its length can be reduced to 6-7m and the total reduction and number of passes is increased.

It was assumed for the calculation that:

1) the time taken to feed a tube into the die and grip it, $t_{gr} = 0.5min$
2) the inertia travel of the carriage $\ell_{in} = 5m$;
3) the carriage acceleration and braking travels $\ell_{acc} = \ell_{br} = 3m$;
4) the rate of acceleration of the carriage at the start of drawing $v'_d = 0.5v_d = 60m/min$
5) the mean velocity when the carriage is retarded while returning $v'_{ret} = 40m/min$

The total drawing time for one 8 x 0.5mm tube in six passes on a drawing block, using a 30 x 1.5mm blank ($\mu_\Sigma = 11.4$; $\mu_{mean} = 1.5$) is determined using equation (107)

$$t_\Sigma = \frac{50}{500} \cdot \frac{1.5(1.5^6 - 1)}{1.5 - 1} + 6 \cdot 1.0 = 3.12 + 6 = 9.1min$$

The output of a block of metres of finished tubes is

$$M = \frac{60}{9.1} \cdot 50 \cdot 11.4 = 3760m/h$$

Obviously, with this schedule, one billet mill can serve $n = 9.1 : (5.14 : 3) \approx 5$ blocks, or if only one block producing 8 x 0.5mm tubes is operated the mill will only be ~20% utilized.

When tubes are produced by a "cold rolling - drawing" schedule, the tube cold rolling mill is the plant producing the blanks. With a linear displacement $m\mu_\Sigma = 45mm$ in each double mill stand pass, the output of a KhPt-12-20 three-stand mill (VNIIMETMASh), at 120 passes a minute, will be

$$M = \frac{3 \cdot m\mu_\Sigma \cdot 60 \cdot n \cdot k}{1000} = \frac{3 \cdot 60 \cdot 60 \cdot 120 \cdot 1.0}{1000} \approx 1300m/h^*$$

The rolling can be carried out to a grooving of 30 x 2.5 - 16 x 0.5 with a total reduction $\mu_\Sigma \simeq 12$. With a finished tube length of 25m 8 x 0.5mm tubes are drawn in two passes ($\mu_\Sigma = 2.07$) with a mean reduction $\mu_{mean} = 1.44$.

In this case it is effective to use drawing blocks with a drawing speed of 50-100m/min positioned beyond the tube cold rolling mill. With $v = 100m/min$ and $t_{aux} = 0.5min$ (drawing without a mandrel - t_{aux} is therefore lower than in the previous case).

* *In this calculation the "strand" coefficient was taken as being 3. the number of strands in a mill. It was found, however, when a KhPT-12-20 mill was used, that this coefficient was 2.1-2.3*

$$t_{\Sigma} = \frac{25}{100} \cdot \frac{1.44(1.44^2-1)}{1.43 - 1} + 0.5 \cdot 2 = 1.88min$$

Drawing block output, in metres of finished tube,

$$M = \frac{60}{1.88} \cdot 25 \cdot 2.07 = 1660m/h$$

In order to compare the technological schedule considered, the consumptions of machine-hours (cold production) per 1000m of finished tubes are given in Table 47. The adjustment coefficients for cold processing are taken as being 1.0. This is approximately equivalent to actual fact, since when long tubes are coil drawn or rolled in tube cold rolling mills the wastage is extremely small.

Table 47

Machine-hours Consumed per 1000m of finished 8 x 0.5mm Copper tubes Produced by Cold Processes

Equipment	Schedule Nos.		
	I	II	III
Preparation plant:			
three-strand draw bench	-	0.049	-
tube cold rolling mill	-	-	0.372
Drawing block	0.167	0.266	0.602
Totals	0.167	0.315	0.974

It should be noted, in conclusion, that when small diameter tubes are produced, in designing new shops or reconstructing existing shops the schedule selected is governed by the production floor area and the range of products. For instance the area required for a tube cold rolling mill and drawing block (schedule III) is more than 80% less than the floor area required for a "Chevalier" type draw bench and a high-speed drawing block (schedule II). Schedule II is, however, more versatile owing to the possibility of using a draw bench for producing a wide range of tubes. Schedule I requires additional floor space at the forward end of the press, and is only rational for producing copper tubes, when no intermediate annealing is required. In addition to this, when copper tubes are produced by schedule I, but not in coils, it is necessary to have a special straightening machine and saw beyond the drawing blocks.

When small diameter copper tubes are produced, the advisability of including preparatory chain-driven draw benches is governed, among other things by the power of the presses available.

EVALUATING THE ECONOMIC EFFECTIVENESS OF PRODUCING 6 x 5(0.5)mm TUBES BY DIFFERENT TECHNOLOGIES

Let us consider the following schedules:

schedule I - 3150t press - 75t chain draw bench - 50t chain draw bench - 1500mm dia. drawing blocks with a drawing force of 8.6t and 1000mm dia. drawing blocks with a force of 2.5t

schedule II - 3150t press - 1500 and 1000mm dia drawing blocks
schedule III - 3150t press - 50t chain draw bench - 1500 and 1000mm dia drawing blocks.

When the rational dimensions of the billets were determined, the basis used was that the maximum extruding force $P_{max} \leqslant 0.85P_{nom} = 2700t$. The dimensions of the billet and the extruding force were determined by the procedure set forth in Chapter I.

With schedule I, tube blanks with dimensions of 72 x 60 (6)mm are extruded from 290 x 490mm billets (D_k = 300mm). With a useful length for the 50t draw bench of 45m, three blanks 7350mm long are cut from the extruded tube. With this variation tubes are drawn in two passes in the 75t draw bench down to 48 x 41.5 (3.25)mm, and in two passes in the 50t draw bench down to 34 x 30 (2.0)mm. The tubes are then drawn in five passes on the 1500mm dia. drawing block down to 13.7 x 12 (0.85)mm and in four passes on the 1000mm dia. drawing block. After drawing on the 1500mm dia. drawing block, the coil is cut in half (the length of the 1000mm dia. drawing block is about 1m)

The additional schedule Ia is provided in order to increase the output of the 3150t press; with this schedule drawing is carried out using thicker walled blanks than with schedule I. The 90 x 75 (7.5)mm blanks are extruded from 290 x 655mm billets. The length of the extruded blanks (not allowing for the discard) is 18800mm, and the tube number factor $n = 4$(as against $n = 3$ with schedule I). The extruded blanks are cut into two halves, each 9400mm long, and transported to the 75t draw bench.

The dimensions of the tubes after the first pass are 72x60 (6.0) x 14600mm, μ = 1.56, and the two-strand drawing force is 72.0t. In the second pass, tubes are drawn down to 58 x 49 (4.5) x 24100mm, μ = 1.64. Though the reduction in this pass is the same as that in schedule I, the tube is already work-hardened, and in this case tubes can only be drawn in two strands with a force of 74.2t. After this pass, tubes are cut into halves 12 050mm long, and a third pass is carried out in the same draw bench, down to 48 x 41.5 (3.25) x 20 000mm, with a coefficient of reduction μ = 1.66. The drawing force in the three strands is 74.1t

The schedule for further drawing in the 50t draw bench and on the drawing blocks is exactly the same as with schedule I. With schedule II, 40 x 35mm extruded blanks are drawn on blocks. The optimum container diameter D_k for extruding these blanks in a 3150t press is 250mm (245 x 360mm billet). Drawing 6 x 0.5mm tubes from thicker walled blanks (for instance t_{bl} = 3mm) requires a greater number of passes and additional diametral reduction, since drawing is carried out on a floating mandrel. This increases the initial diameter of the blank to 50-55mm. It is difficult to draw tubes using these blanks on blocks with a diameter of 1500mm.

Allowing for a discard of 1000m from the front and rear ends, the length of the extruded blanks is 50 000mm. As with the previous schedule, the blanks are extruded into water without a subsequent pickling operation. The blanks thus produced are coiled and transported to the 1500mm dia. drawing blocks. Drawing is carried out in ten passes: 6 passes on the 1500mm block and 4 passes on the 1000mm block. Except for the first of these, the drawing schedules are the same as with technological schedule I. Schedule II envisages the intermediate cutting of tubes after the sixth pass.

With schedule III, blanks with dimensions of 80 x 70 (5.0)mm are extruded from 240 x 480mm billets (D_k = 300mm), and are taken to the 50t draw bench, where they are drawn in four passes down to 40 x 35 (2.5)mm. From this point the ten-pass block drawing is equivalent to schedule II

The length of the extruded tube is 23 000mm. After deducting a discard of 600mm, this corresponds to two blanks for drawings each about 11 200mm long. After the fourth pass in the 50t draw bench the length of the tube is 45 000mm.

With all the schedules (except for Ia), the tubes are drawn in three strands on the chain draw benches. The billets are heated to 920-950°C before extrusion.

The optimum process schedule must provide the following:

1. A combination of extrusion, drawing in three-strand draw benches and drawing on 1500 and 1000mm dia. drawing drums, with which the maximum use is made of the output of each plant separately, with

outputs interconnected in each of the operations.

2. Minimum delays in production under particular conditions.

3. Minimum capital outlay and the maximum performance by the equipment used.

The above conditions govern the selections of the technical and economic figures by which the proposed variations are compared.

a) output of the principal plant and the load on it;
b) cost of production and outlay on extrusion and drawing
c) specific capital outlays and reduced outlays for all the variations considered.

Then the output of the principal plants were calculated, the following figures were used as bases; the 3150t press-produces 32 extruded billets an hour; the drawing rate for the 75t draw bench is 75m/min, that of the 50t draw bench 100m/min, the drawing speeds on the blocks 300m/min. Details of the outputs of the principal plant operating by the different technological schedules are given in Tables 48-50.

Table 48

Outputs of the 3150t press

Schedule	Mass of blank kg	Press hourly output t	
		of billets	of extruded blanks
I	244	9.22	7.66
Ia	325	12.28	10.40
II	133	4.83	4.26
III	236	9.02	7.55

The comparison of schedules I, Ia, II and III by the output of the equipment makes it possible to determine the process schedules which utilize the equipment concerned most rationally. Analysis of the output of the 3150t press (table 48) shows that the press can be used most effectively with schedule Ia. This schedule uses billets with $L_b = 655mm$, and the extruded blanks have a wall thickness $t_{bl} = 7.5mm$. With schedules I and III press output is about the same in each case. The output of the 3150t press is lowest with schedule II, extruding billets with $L_b = 345mm$ into blanks with wall $t_{bl} = 2.5mm$ thick.

Table 49

Outputs of the 75 and 50t draw benches

Schedule	Draw bench force	pass nos	Duration of drawing cycle per tube h	Draw bench output	
				m/h	t/h
1	75	1	0.1195	6050	40.8
		2	0.1668	7200	29.4
		Overall to the prescribed dimensions	0.2863	4200	17.1
	50	1	0.1988	9300	24.4
		2	0.2731	3940	17.8
		Overall to the prescribed dimensions	0.4719	5740	10.3
1a	75	1	0.1001	8 800	97.6
		2	0.1445	10 000	67.5
		3	0.1668	7 200	29.4
		Overall to the prescribed dimensions	0.4114	2 990	11.8
	50	1	0.1988	9300	24.4
		2	0.2731	9940	17.8
		Overall to the prescribed dimensions	0.4719	5740	10.3
III	50	1	0.1177	7620	60.00
		2	0.1456	8150	48.50
		3	0.1915	9220	36.90
		4	0.2725	9900	26.00
		Overall to the prescribed dimensions	0.7273	3710	9.72

With schedule I the output of the 75t draw bench is 66% higher than that of the 50t draw bench i.e. the 75t draw bench will not be so heavily loaded in operation as the 50t draw bench (table 49). The lower output when the 75t draw bench is used with schedule Ia is due to the fact that, at the same time as the drawn blank number factor for the first two passes is increased, the number of strands decreases from 3 to 2. Owing to drawing heavier section blanks, the output from the 75t draw bench is lower.

Schedule III is such that the extruded blank is taken immediately to the 50t draw bench. In this case the output of the draw bench is slightly lower than with schedules I and Ia, because the blanks are larger and the number of passes is greater. Schedule III, however, utilizes the 50t draw bench better, and there is no need for the costly three-strand 75t draw benches.

When the output of the stage of drawing with drawing blocks was calculated, the non-overlapping auxiliary time was taken as being $t_{aux} = 45s$ for all the passes. The speed of the drawing blocks with a diameter of 1500mm in the first three passes was 100m/min, while in all the subsequent passes (in all the passes in the case of the 1000mm dia. blocks) it was 300m/min; this was in accordance with the technical particulars of the drawing blocks and the actual running data. Calculated drawing outputs are given in table 50. The slight fluctuations in output from the 1500 and 1000mm dia. drawing blocks with the schedules compared are due to slight changes in the length and wall thickness of the blanks reaching the drawing blocks, while the drawing speeds and coefficients of reduction for the passes are constant.

Calculation of the Utilization of Equipment and the amount required. Calculation of the hourly output figures for equipment used with the different technologies compared was the basis on which the standard (norm) machine hours per ton of tubes, also the amount of equipment required and its utilization, were determined. The standard machine-hours obtained were adjusted by the coefficient of operative time for the type of equipment concerned; this coefficient allows for preparatory and finishing times and the durations of regular stoppages; the norm machine-hours were also adjusted for the metal consumption coefficient. The values of machine-hours per ton of finished tubes are given in Table 51.

The amounts of equipment required were calculated for the same conditional programme of producing 30 000t of finished tubes a year. When the available useful working time was calculated for equipment, allowance was made for stoppages due to running repairs to the equipment according to the planned preventative repairs (PPR) chart; stoppages due to changing tools were also taken into account. The results of calculating amounts of types of equipment and their utilization coefficients are given in Table 51.

Table 50

Drawing blocks outputs $\frac{m/h}{t/h}$

Schedule	Pass nos 1	2	3	4	5	6	Total for drawing down to the final dimensions	8	9	10	11	Total for drawing down to the final dimensions
	1500mm dia. drawing block							*1000mm dia. drawing block*				
I	2350/3.390	3360/2.940	6850/3.560	8300/3.490	9700/2.970	–	2160/0.663	8600/1.800	9950/1.434	11510/1.160	12580/0.980	4110/0.317
II	2960/5.300	3560/4.240	4050/3.420	9000/4.680	10480/4.400	11780/3.600	2310/0.709	7360/1.540	9000/1.298	10620/1.070	11740/0.905	3760/0.290
III	2800/5.100	3420/4.060	3900/3.290	8500/4.420	10000/4.420	11400/3.490	2210/0.676	8200/1.715	9840/1.420	11450/1.150	12500/0.964	4080/0.314

Table 51

Machine hours norms, amount of equipment and coefficients of its utilization

Equipment	Schedule	Norm machine-hours per ton of blanks	Metal consumption coefficient	Norm machine-hours per ton of finished tubes	Calculated available machine-hrs. for finished tube prod.	Number of units of equipment		Coefficient of utilization of equipment
						cal-culated	Inst-alled	
3150t press, K*= 0.97 ϕ** = 5250 h	1	0.1310	1.240	0.167	5000	0.95	1	0.950
	1a	0.0963	1.235	0.123	3690	0.705	1	0.705
	11	0.2350	1.145	0.277	8300	1.58	2	0.790
	111	0.1320	1.238	0.168	5050	0.97	2	0.490
75t draw bench K = 0.80 ϕ = 5500 h	1	0.058	1.050	0.072	2145	0.39	1	0.390
	1a	0.085	1.045	0.105	3150	0.58	1	0.580
50t draw bench K = 0.80 ϕ 5500 h	1 & 1a	0.097	1.020	0.124	3720	0.68	1	0.680
	111	0.103	1.037	0.134	4020	0.73	1	0.730
1500mm dia. drawing block, K = 0.91 ϕ = 5378 h	1 & 1a	1.510	1.010	1.672	50 160	9.35	10	0.935
	11	1.410	1.012	1.572	47 160	8.80	10	0.880
	111	1.480	1.010	1.640	49 200	9.15	10	0.925
1000mm dia. drawing block, K = 0.91 ϕ = 5378 h	1 & 1a	3.155	1.010	3.470	104100	19.40	21	0.925
	11	3.450	1.010	3.800	114000	21.20	23	0.925
	111	3.183	1.010	3.520	105600	19.70	21	0.900

* Coefficient allowing for time consumed on preparatory and finishing operations, and regular stoppages

** Useful available equipment working time in the year

Comparison of the utilization figures for equipment shows that, with the accepted optimum technological parameters for processes, with schedules I and Ia, one 3150t press can cover the cold processing demand for tube blanks. With the number of extrusions an hour used in the calculation, one press can provide the annual programme by schedule III, but with a high coefficient of utilization (0.97). Since the calculation was made for a conditiona standard dimension, and the probability of error can approximately be taken as 5%, the permissible utilization coefficient was taken as being not over 0.95. For this reason, the figure for the number of presses for this schedule had to be taken as 2.

The production of tubes by schedule II, which only involves the use of presses and drawing blocks, also requires two presses, since the blanks used with this technology are thin-walled. With schedule I, Ia and III, based on the use of line draw benches, the calculation shows that one 75t draw bench and one 50t draw bench are required.

With all the technologies, about the same number of 1500 and 1000mm dia. drawing blocks are required. The reason for this is that, as was said earlier, all these variations are based on the same drawing block output. In addition to this the equal number of drawing blocks for all the schedules makes it possible to solve the problem of whether it is advisable to use 75 and 50t draw benche for producing small diameter copper tubes, also to select the optimum schedule observing the conditions for their comparability.

Comparative Assessment of Different Technological Schedules

The comparison and comparative evaluation of the economic effecti ness of different technological schedules were carried out first i terms of the cost of a ton of finished tubes, and included the following expenditure headings:

1) the cost of metal after deducting wastage
2) the pay of the principal operators
3) expenditure on electric power for motors
4) expenditure associated with the running of equipment and the flo space occupied by it.

Since the principal technological operations in the process (extr sion and drawing) were considered, the schedules were only in term of the above headings.

The expenditure on electric power for motors was determined on th basis of the motor power installed, and the coefficients of utiliza tion of the motors by time and power. The expenditure involved in the utilization of production floor space was calculated because it varies greatly according to the type of equipment installed.

Schedule II provides the lowest metal consumption coefficients (Table 52) and consequently the lowest cost after deducting wastage, by comparison with schedules I, Ia and III; this is due to the smaller wastage of metal on plugging heads when drawing is carried out with the 75 and 50t draw benches.

Table 52

Yields of Sound Metal and Metal Consumption coefficients for the schedules compared

Indices	Technological schedules			
	I	Ia	II	III
Yield of sound metal % of billet	80.50	80.80	87.35	80.70
Metal consumption coefficient	1.24	1.235	1.145	1.238

Details of the cost of a ton of finished tubes are given in Table 53. The figures in Table 53 show that the running costs are about the same for process technologies I, II and III, comprising 42.20-42.50 roubles per ton of finished tubes. The reason for this is that the expenditures under the individual headings are different for the variations considered but total to about the same. For instance, owing to the higher output of the 3150t press, by comparison with schedule II, schedules I and III provide an overall reduction in work per ton of finished tubes, and consequently lower figures for pay and cost of electric power for the motors.

The use of 75 and 50t draw benches, however, tends to increase the outlay associated with running the equipment and utilizing floor area. The use of thin walled blanks with schedule II increased the expenditure on their extrusion, but since the production process does not include costly draw benches, even when two presses are used there is a slight reduction in the outlay on running the equipment.

The fact that the outlay on the entire process is lower (41.39 roubles) with schedule Ia is because the press is better utilized than with any of the other technologies. Comparison of all the outlay figures allowing for the cost of the metal after deducting wastage shows that schedule II provides the lowest cost figure (745.68 roubles) owing to the lower metal consumption coefficient. The cost figures for a ton of finished tubes produced by schedules I, Ia and III are about the same.

Table 53

Calculated Costs, specific capital investments and reduced outlay figures for a ton of finished tubes produced by the different technologies

Items	Technological schedules I	Ia	II	III
Cost of processing a ton of finished tubes roubles				
cost of metal after deducting wastage	709.2	709.0	703.6	709.0
expenditure on processing	42.51	41.39	42.18	42.21
This includes:				
operators' pay	6.86	6.77	7.04	6.77
depreciation on the value of :				
equipment	16.22	16.22	15.31	16.12
floor area	1.50	1.50	1.18	1.38
expenditure on repairs and maintenance:				
equipment	9.58	9.58	9.27	9.22
buildings	0.55	0.55	0.43	0.50
expenditure on electric power	7.80	6.77	8.95	7.52
TOTALS	751.71	750.39	745.68	751.21
Specific capital investment per ton of tubes roubles	211.91	211.91	197.55	212.52
Reduced outlay per ton of tubes roubles	783.46	782.14	775.28	783.01
Work per ton of finished tubes:				
machine hours	5.505	5.493	5.649	5.452
man-hours	12.122	11.957	12.406	11.978

If schedules I-III are compared by production costs alone, we find that the minimum expenditure is provided by changing to schedule II. In addition to this, however, the schedules were also compared by specific capital outlays and reduced figures (see Table 53). With schedules I, Ia, and III, the specific capital outlays and reduced outlay figures were about the same (respectively 212.0-212.5 roubles and 782-783 roubles per ton of finished tubes). Consequently the technical schedules including the use of 75t and 50t draw benches are equally as effective from the economic point of view.

The specific outlay figure is lowest for schedule II. By comparison with schedules I, Ia and III, with schedule II the reduced expenditure figures are about 7.20 roubles lower per ton of finished tubes. The use of drawing blocks alone, in the production line, after the 3150t presses, for producing small diameter copper tubes thus provides the minium production costs, also the minimum reduced outlays, and also utilizes the basic equipment best.

Relationship of the Effectiveness of Schedules to press power

Nine methods of producing copper tubes with the same dimensions of 6 x 5mm, involving the use of presses with different powers, were considered in addition to the technologies described above. This comparison was necessary becauses a wide range of press equipment is now being used both at home and abroad for processing non-ferrous metals and alloys. It will be proved that the economic effectiveness of a particular process schedule depends on the power of the press used.

Processing schedules envisaging the use of 1500, 4500 and 8000t presses, other processes being the same as in schedules I-III, were considered. The schedules envisaging the use of 4500 and 8000t presses included drawing on 2100mm dia. drawing blocks. The technological parameters of the processes were calculated, for the additional variation, by the procedure described earlier.

In order to make all the technical and economic indices comparable for the 4500 and 8000t presses, the dimensions of the blanks were taken as being the same as with the schedules considered earlier. The output of the press equipment was taken, in accordance with its technical particulars, as being 60 extrusions an hour for a 1500t press, 28 extrusions an hour for a 4500t press, and 18 extrusions an hour for an 8000t press.

In the schedule including an 8000t press and drawing blocks, the blank dimensions were taken as 61 x 55mm; this meant that the number of passes on the 2100mm dia. block was increased to seven as against the six passes with the equivalent variations. As regards the other process schedules, the total numbers of passes on the draw benches and drawing blocks were not changed. Equivalent technical and economic comparisons were made for the additional process schedule The calculations were based on the same assumptions as for the previous schedules. The results of comparing all the different technological schedules are presented diagrammatically in fig.67

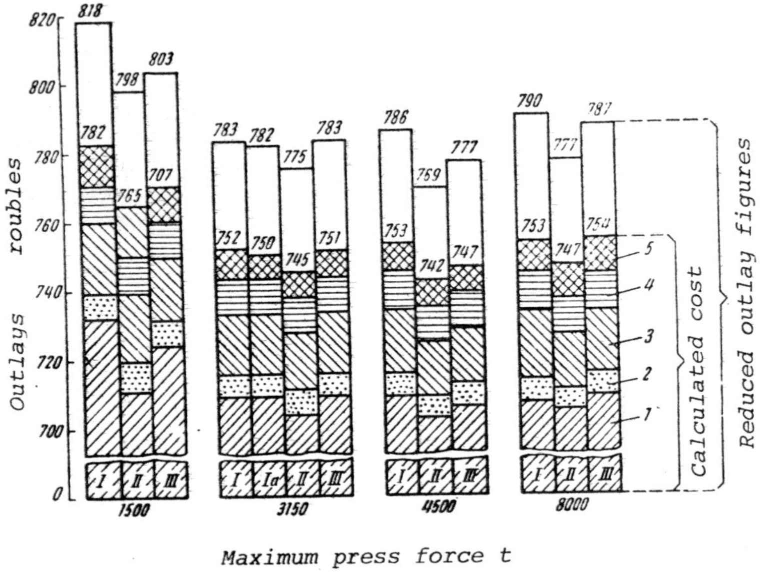

Fig.67: *Diagram of the calculated cost and reduced outlay figures for a ton of finished 6 x 5mm copper tubes (I, Ia, II, III a the different technological schedules). Items of expenditu*

1) cost of the metal after deducting wastage; 2) pay; 3) depreciation deductions from the cost of the equipment and shop floor area; 4) outlay on repairs to equipment and buildings; 5) cost of electric power

The schedules including an 1500t press have the highest metal consumption coefficients (1.26-1.60); this is due to the reduction in billet volume during extrusion and the resultant decrease in the number of extruded blanks. For the remaining variations, the yield of sound metal from billets only fluctuates slightly (1.12-1.24). The schedules which only include drawing blocks provide the highest yield of sound metal.

Comparison of the hourly outputs of presses shows that, with the accepted number of extrusions an hour and the specified blank dimensions, the increase in output on changing from a 1500t press to a 4500t press is 79-105%. The fact that the output of the 8000t press is only slightly higher than that of the 4500t press (only 3% higher) is due to the fact that increase in billet volume is limited by the tube blank dimensions adopted for the schedules considered. The use of thin walled blanks and smaller volume billets in the schedules including drawing blocks in every case reduces the output of the press equipment.

When the press power is increased from 1500 to 8000t, the schedules including 75 and 50t draw benches have the highest figures for cost, specific capital outlay and reduced expenditure. These figures are slightly better for the variations not including 75t draw benches. The highest costs and reduced expenditure figures are those for the schedules with 1500t presses. When 3150 or 4500t presses are used, these figures are lower, while they increase slightly when 8000t presses are used, (see fig.67).

The variations in costs and reduced outlay figures with the technologies entirely eliminating chain draw benches, including only block type drawing equipment, are extremely interesting. These variations provide the lowest production outlays and specific capital outlays. The reason for this is that costly draw benches are not needed, the floor area occupied by them is not required, and the outlay involved in running them is reduced. By comparison with schedules envisaging the use of 75 and 50t draw benches, the reduced outlay figures for these variations are 1.5-3% lower.

With regard, therefore, to all the technologies for the production of small diameter copper tubes which have been considered, schedules only including drawing blocks are the most economical and technically justifiable. It is not, however, rational to use 1500 or 8000t presses for producing 30 000t of small diameter copper tubes per annum, since the 1500t presses have low outputs and the high cost of 8000t presses increases the capital outlay and the costs of running them. The rational equipment for different annual productions will not be the same.

PRODUCTION OF GERMAN SILVER TUBES

Various technologies are employed at present in the production of German silver tubes (MNZhMts 30-0.8-1.0 alloys). Let us consider the rationality of technological schedules using the production of German silver tubes with dimensions of 16 x 13 (1.5) x 2740mm as the example.

Schedule I: Solid blanks with dimensions of 88 x 3520mm, are extruded in a 3000t horizontal hydraulic press. These are then cut, drilled, end machined and the heads countersunk to form blanks for the 600t vertical hydraulic press. The extruded 32 x 26 (3.0) x 2040mm extruded blanks are subjected to preparatory operations (scraping, pickling, washing and forging the grips), and then drawn in four passes in a 5t single strand draw bench. The fifth and sixth passes are carried out in a single-strand 3t draw bench. This technology, used at one of our works, suffers from a number of failings: the yield of sound metal is low (44.32%) there are many operations, and the production cycle is very long. Process schedules eliminating double extrusion were developed in order to improve the yield of sound tubes and reduce the technological costs.

Calculations have proved that schedules for extruding blanks in a 1500t press and then rolling them in a tube cold rolling mill are not effective: when blanks with walls 5mm thick are extruded from a container with a diameter of 155mm it is impossible to ensure consistent extrusion owing to the insufficient power of the press.

Schedule II: Blanks with dimensions of 55 x 38 (7.0) x 6560mm are extruded in a 3000t press (200 x 250mm billets). The blanks are then rolled, in a KhPT-55 cold rolling mill, down to 22 x 18.4 (1.8) x 3700mm and drawn down to their final dimensions in two passes.

Schedule III: Billets with dimensions of 200 x 265mm are extruded in a 3000t press into blanks with dimensions of 60 x 45 (7.5) x 5500mm, which are rolled in a KhPT-75 cold rolling mill into 38 x 32 (3) x 3710mm tubes, then in a KhPT-32 cold rolling mill into 20.0 x 16.8 (1.6) x 4400mm tubes. This is followed by a single sizing drawing pass.

The length of the discard after extrusion with the technologies compared was taken as being about 800-900mm in every case. When the outputs of the cold rolling mills were calculated, the basis used was the linear displacement $m\mu_{\Sigma}$: for the KhPT-75 mill - 75mm (output 211m/h); for the KhPT-55 mill - 52mm (214.5m/h); for the KhPT-32 mill - 43mm (190m/h). With all the process schedules, the consumption coefficients were only determined allowing for the commercial wastage.

Table 54

Consumption coefficients and costs of metal for the different technological schedules for producing 16 x 13 (1.5) x 2740mm tubes of MNZhMts 30-0.8-1.0 alloy

Items	Units of measurement	Schedule nos.		
		I	II	III
Yield of sound metal	kg	44.32	71.40	68.12
Overall consumption coefficient (1500-3000t presses)	kg	2260	1400	1468
Cutting consumption coefficient	kg	2115	1160	1212
Extrusion consumption coefficient (600t press)	kg	1485	—	—
Rolling consumption coefficient	kg	—	1065	1080
Drawing consumption coefficient	kg	1390* 1240**	1065	1080
Cost of the metal	roubles	2971-74	1840-90	1930-31
Amount of wastage	kg	1260	400	468
Cost of wastage	roubles	1386-00	440-00	514-80
Cost of metal after deducting cost of wastage	roubles	1585-74	1400-90	1415-51

* First two drawing passes

** Subsequent drawing passes.

Table 54 gives the consumption coefficients and the cost of the metal after deducting wastage for the schedules compared. Comparison of the consumption coefficients convincingly indicated the great advantage possessed by producing tubes by cold rolling. By comparison with schedule I (44.32% yield of sound metal), the highest yield is provided by schedule II (71.40%). The considerable increase in the yield of sound metal with schedules II and III is due to the fact that replacing double extrusion by cold rolling makes it possible to eliminate the machining of billets before they are extruded in the vertical 600t hydraulic press, also to eliminate the considerable losses of metal associated with this, and the consumption of metal inevitable in several drawing passes is also reduced. In addition to this, by comparison with schedule I, schedules II and III greatly reduce the work consumption— the man-hours and machine-hours are more than halved.

It can be seen from the figures in Table 55, which give the items of expenditure on producing a ton of finished tubes, that the process cost of producing tubes by schedules II and III is 12-15% lower than with schedule I. The lower technological cost is basically due to the lower consumption coefficient the lower expenditure on metal, and the lower pay figures. The increase in depreciation due to the cost of the equipment and the increase in expenditure on running repairs to equipment and its maintenance are due to the higher cost of the rolling mills which replace the cheaper 600t press and draw benches.

*　　*
*

Only certain process schedules and their effectiveness by comparison with others have been considered in this chapter. The authors have not attempted fully to consider the wide range of possible technologies. The schedules considered make it possible to define the possibilities of each process (extrusion, drawing and cold rolling) individually. The details given in this chapter can be used in developing other schedules, and make it possible to evaluate whether particular technological schedules are practicable.

When, however, individual technological schedules for producing a particular range of tubes are compared apart from the shop programme and range of products subdivided by month or year, it is impossible to establish whether any particular schedule is rational or to give preference to a partiuclar schedule. A schedule rational under certain conditions will not be rational under others.

Table 55

Work Required for Producing 16 x 13 (1.5) x 2740mm tubes of MNZhMts 30-0.8-1.0 Alloy Tubes by the different technological schedules I-III

Operations	Schedule Nos. I		II		III	
	Machine hours	Man hours	Machine-hours	Man hours	Machine hours	Man hours
Extrusion	0.469	3.283	0.750	5.290	0.763	5.341
Straightening of blanks	0.516	1.032	0.378	0.378	0.378	0.378
Machining of billets	27.610	27.610	-	-	-	-
Extrusion in 600t press	4.550	18.200	-	-	-	-
Cutting of ends and removal of fins, pickling and scraping	12.605	14.550	1.613	1.996	1.632	2.010
Rolling in different mills:						
KhPT-75	-	-	-	-	2.060	2.060
KhPT-55	-	-	5.730	5.730	-	-
KhPT-32	-	-	-	-	8.050	8.050
Forging grips	1.085	1.085	1.164	1.164	1.180	1.180
First two drawing passes	3.785	3.785	-	-	-	-
Removal of grips	0.748	0.748	-	-	-	-
Rolling of grips	3.903	3.903	-	-	-	-
Drawing	7.514	7.514	4.200	4.200	2.310	2.310
Finishing operations	26.628	45.826	25.954	44.806	26.062	44.944
Totals	89.413	127.536	39.789	63.561	42.435	66.273
Total machine-hours without manual operations	76.810		34.049		36.681	

At the same time it is extremely important to find the most rational variations within the particular technological schedules. This is convincingly proved by the example of producing MNZh5-1 alloy tubes. The effectiveness of the schedule can be greatly improved merely by altering the dimensions of the blank and the number of tubes per blank in the extrusion-drawing schedules.

The procedure for selecting the parameters of the process in order to establish the optimum variation of a particular technology has been considered for each of the processes in the respective chapters. The task of finding the most rational schedule for producing individual types of tube must thus be solved allowing for the volume of production from the shop concerned.

CHAPTER V SELECTION OF RATIONAL PROCESS SCHEDULES

At present there is the following sequence of utilization of equipment in most tube drawing shops. On the basis of the quarterly plan for production (subdivided by months) the senior planner in the shop calculates the production, for the ten day period subdivided into days, of all alloys and all sections and dimensions. On the basis of these data, the works production department compiles, and authorizes, a schedule for the supply of billets for the ten day period, subdivided into days. The extrusion press department planner takes into account the billets left over from previous days and the supply of billets in the current day and plans the daily utilization of the equipment in the press shop and the production of extruded blanks and finished products.

Using the plan for the daily production of extruded blanks, the planner in the drawing department determines the daily production of drawn and rolled products and compiles the orders for the shop shift despatcher; these indicate the component (section and size, alloy, daily production), the blank, and the technological schedule. It is important to emphasize that the distribution by types of equipment, i.e. the daily utilization of equipment in the drawing department, is performed by the despatcher on the basis of the existing process schedules and his personal experience.

In many cases the fabrication techniques do not include a sufficiently scientifically based technical and economic appraisal for the different particular production conditions, and they may not always be rational (optimal) under particular conditions. It should be noted that, in shop conditions, it is possible to produce many drawn and rolled products by different sequencies. There are different accepted criteria for evaluating the effectiveness of production methods. Some authors (91) consider that the correction co-efficients (consumption of metals) represent this criterion, others (92,70,9) that it is the production cost of products or specific technical and economic indices: consumption of machine-hours or man-hours pay, expenditure on metal and electric power, etc.

PRODUCTION METHODS FOR BRASS CONDENSER TUBES

As an example, we will consider the production of 28 x 1.0mm L68 alloy condenser tubes. Various technologies are used for producing these tubes (Table 56): dual extrusion and drawing (schedule I); extrusion, cold rolling and drawing (schedule III, VI and VII); extrusion and drawing (schedule VIII).

Table 56

Technology variations for the production of 28 x 1.0mm L68 Alloy Condenser Tubes

Equipment	Schedules							
	I	II	III	IV	V	VI	VII	VIII
Extrusion								
3000t press	250x750* / 97.5	-	-	200x290 / 65x7.5	200x290 / 65x7.5	200x280 / 60x5.0	200x280 / 60x5.0	-
1500t press	-	150x230 / 38x4.0	150x240 / 38x4.0	-	-	-	-	150x265 / 46x3.0
600t press	97.5x138* / 39x2.0	-	-	-	-	-	-	-
Cold Rolling								
KhPT-75 mill	-	-	-	38x3.5		32x1.5	-	-
KhPT-55 mill	-	-	-	-	38x3.5	-	32x1.5	-
KhPT-32 mill	-	28x1.0**	30x1.2	28x1.0	28x1.0	-	-	-
Drawing ("Spidem" draw bench)								
Passes:								
1	33.6x1.8**	-	28x1.0	-	-	30.4x1.2	30.4x1.2	42.2x2.6
2	32.0x1.5	-	-	-	-	28.0x1.0	28.0x1.0	38.8x2.3
3	30.4x1.2	-	-	-	-	-	-	36.0x2.0
4	28.0x1.0	-	-	-	-	-	-	33.6x1.8
5	-	-	-	-	-	-	-	32.0x1.5
6		-	-	-	-	-	-	30.4x1.2
7	-	-	-	-	-	-	-	28.0x1.0

*The upper figures are the billet dimensions (diameter and length), and the lower figures are the dimensions of the extruded blank (external diameter, wall thickness), in mm

The basis upon which tee output of tube cold rolling mills was calculated in reference (13) was their technical characteristics and the linear movement of the metal; for the KhPT-32 mill, $m\mu_{\Sigma}$ = 45mm, for the KhPT-55 mill $m\mu_{\Sigma}$ = 60mm, and for the KhPT-75 mill $m\mu_{\Sigma}$ = 71-75mm. The data given in reference (70) were used as the basis for calculating the output of the "Spidem" draw bench. In many cases the bench force (15t) was the limiting factor, and drawing was carried out in one or two strands (the first three passes in schedule VIII). After rolling in the tube cold rolling mills (schedules III, VI, and VII), the schedule envisaged intermediate annealing in electric furnaces. In schedule VIII, after the third pass the tubes were cut in half and subjected to an intermediate anneal.

Let us condider the technical and economic figures for the technology variation given in Table 57.

Yield of sound metal: (determined from the geometrical dimensions of the wastage). All the schedules, except for the extrusion of blanks in a vertical 600t press, have higher yields of sound metal (compare schedules I and II-VIII). When cold rolling is used the consumption of metal increases from 63.43% (schedule I) to 78.65% (schedules VI and VII).

The labour in man-hours: for schedules VI and VII, in which KhPT-75 and KhPT-55 mills are the equipment used, is about 30% lower than for schedules I and VIII. The amounts of machine-hours are about the same for all the schedules, about 10.5-12 machine-hours per ton of finished tubes. The exceptions are schedules V and VI, which include double rolling: here the machine-hours are about 25-30% more than with the other schedules

The expenditures on processing (not taking into account the cost of tools and the electric power consumption in the heat treatment furnaces) is lowest for schedule I. With schedules II-VIII the expenditure on processing is basically higher because of depreciation from the cost of equipment, expenditure on electric power for motors, and running repairs to equipment. The reason why the depreciation is 30-200% higher with schedules II and VII lies in the high cost of the rolling mills and the fact that, in production the debts incured with them are higher than in the case of d draw benches (schedule I). Owing to drawing in one or two strands in the first three passes and to the large number of passes the expenditure on electric power is much higher with schedule VIII. The higher expenditure on running repairs to equipment with schedules I-VII than with schedule VIII is due to the greater amount of work required to repair mills and vertical 600t presses.

Table 57

Technical and economic figures for the production of 28 x 26 x 9000mm L68 Alloy tubes by the different technologies

Items	Technologies							
	I	II	III	IV	V	VI	VII	VIII
Consumption coefficient t/t	1,580	1,286	1,330	1,378	1,378	1,270	1,270	1,475
Cost of the metal after deducting wastage roubles t	606—25	588—34	591—02	593—95	593—95	587—36	587—36	598—85
Pay roubles t	37—09	34—80	36—78	35—00	34—80	33—40	33—60	39—02
this including pay by the normed technological chart	11—97	12—72	12—97	13—12	12—99	12—35	11—33	12—81
Outlay on running equipment roubles t								
depreciation	8—64	20—58	16—56	26—61	24—77	15—25	11—29	10—19
electric power (motors)	14—85	8—55	10—93	13—58	11—53	17—95	13—44	26—24
running repairs to equipment	3—64	12—15	10—01	14—62	14—87	6—30	7—12	1—94
maintenance of equipment	0—88	1—88	1—66	2—44	1—85	1—40	1—04	0—93
Total outlay on processing roubles t	65—10	77—96	75—94	92—25	87—82	74—60	66—49	78—32

Table 57 contd.

Items	Technologies							
	I	II	III	IV	V	VI	VII	VIII
Production cost roubles t	661—35	666—30	686—96	686—20	681—77	661—96	653—85	677—17
Work for the basic production processes: machine-hours/man-hours								
extrusion	2,222 / 9,794	0,747 / 5,229	0,751 / 5,257	0,491 / 3,437	0,491 / 3,437	0,393 / 2,751	0,393 / 2,751	0,755 / 5,285
drawing	1,788 / 3,576	—	0,551 / 1,102	—	—	0,947 / 1,894	0,947 / 1,894	2,979 / 5,858
cold rolling	—	8,000 / 8,000	6,240 / 6,240	10,060 / 10,060	9,920 / 9,920	4,180 / 4,180	4,150 / 4,150	—
auxilliary operations	6,796 / 9,581	3,080 / 4,415	5,196 / 7,285	3,612 / 5,662	3,612 / 5,662	4,955 / 7,554	4,955 / 7,554	7,434 / 12,049
total	10,806 / 22,951	11,827 / 17,644	12,738 / 19,884	14,163 / 19,159	14,023 / 19,019	10,475 / 16,349	10,445 / 16,319	11,168 / 23,192
Production of finished tubes t/h:								
rolling mills	—	0,125	0,160	0,099	1,010	0,239	0,241	—
draw benches	0,560	—	1,810	—	—	1,055	1,055	0,336

The production costs: with schedule VII are 1.2-4.7% lower than with the other schedules owing to reduction in the expenditures on metal and labour. From the point of view of production analysis, schedule VII is economically the best proposition.

This comparison of process schedules is necessary but not sufficient when selecting the equipment for shops being planned. For both shops being planned and shops in operation not one of the technical and economic indices which have been considered is an effective criterion of schedule optimality if a schedule is considered without allowance for the actual conditions. For instance works A uses schedule I for producing the tubes under consideration while owing to the increased plan task and the installation of "Spidem" equipment works B has changed from schedule IV to schedule VII.

In the case of shops being designed, to find the effectiveness of capital outlay when the different schedules are used it is necessary to take into account the complete volume of production the complete range of products, and the proportions occupied by the different types of products in the plan tasks. In the case of existing shops, when the optimum schedules are being selected the available machine-hours for the individual types of equipment, whether spare labour is available or not, also the plan task for the range of products, must be taken into account.

All the process schedules considered above are possible and have been tested in practice over many years. In actual fact, however, the range of schedules is not fully covered by those described, since within each schedule there are different possible ranges of billet and blank dimension, i.e. different distributions of deformation among the types of equipment used for producing particualr items.

The possible variations of producing particular L68 alloy tubes were given earlier. Different variations can similarly be given for different tubes made of other alloys and metals (copper, MNZh5-1 alloy, German silver etc.). Different variations of production using different types of equipment can be found for each tube dimensions. In each case the schedule with the lowest costs can be found for producing tubes of particular dimensions. All the schedules with the lowest cost figures, however, may not in most cases be used for producing tubes of particular dimensions of tubes, owing either to having to use the same equipment and suffering from a shortage of machine hours, or to the duration of the processing cycle and the impossibility, owing to this, of fulfilling the plan; there may be other reasons. Thus the variation with the lowest production cost (this would appear to be the overall index) is not always the optimum variation.

It is extremely difficult to solve a problem with many variables by sorting through all the possible variations, and in some cases it is impossible. The selection of the optimum variations therefore requires a new and improved system of methods and new technical procedures. Linear programming can be used effectively for solving this problem, together with discrete action electronic computers. For this, however, it is necessary to work out mathematical models of tube production.

DETERMINATION OF RATIONAL COMBINATIONS OF DIFFERENT METHODS OF TUBE-MAKING BY MEANS OF A COMPUTERISED MODEL *(developed by A M Rytikov)*

Various methods of tube production are now extensively used: hot rolling and extrusion, cold and warm rolling in KhPT and KhPTR mills, and drawing with or without a mandrel. In most cases the technological schedules are combinations of several of these processes and the sequence of operations is governed by the technical particulars of the equipment, the physical properties of the metal, and economic indices. The dimensions of the blank reaching the cold processing shop, however, also the relationship between the deformations with the different methods, are governed exclusively by the experience of the technologists, and are in many cases far from being optimal.

The range of tubes, made of a particular metal or alloy, which are produced can be represented diagrammatically (fig.68). The smallest tube wall thickness t and internal diameter d are defined by the boundary abc; the smallest blank dimensions $t_{bl(min)}$ and $d_{bl(min)}$ for blanks produced by hot processes are denoted by the boundary de. Thus region I defines the range of tubes produced mainly by hot treatment, and region II the range produced by cold treatment. In many cases, however, when it is required that tubes be produced with strict tolerances on their dimensions and a good surface finish, the cold treatment boundaries are extended.

Each point *M* in the plane of the diagram corresponds to a particular tube, the dimensions *t* and *d* of which are the co-ordinates of that point. Each of the operations is a transition from point $M(d',t')$ to point $N(d,t)$, and can be regarded as the vector $\overline{MN}$. It is easily proved (17) that each *i-th* method of treatment is represented by a quite definite range of deformations along the diameter (Δd_i) and the wall (Δt_i):

$$\frac{\Delta t_i}{\Delta d_i} = k_i \qquad (154)$$

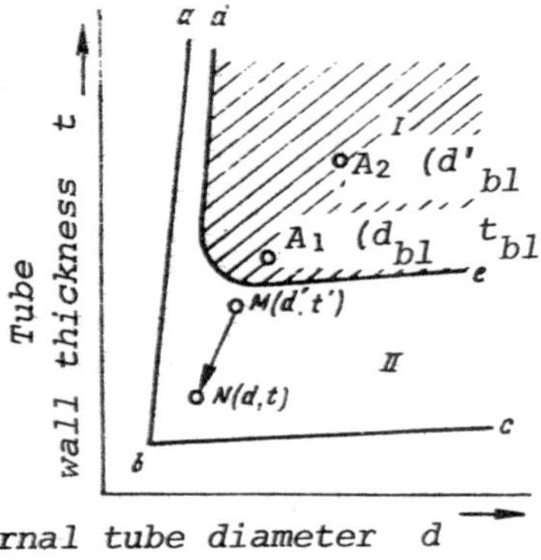

Fig.68: Diagram of the range of tubes: I) hot treatment region; II) cold treatment region

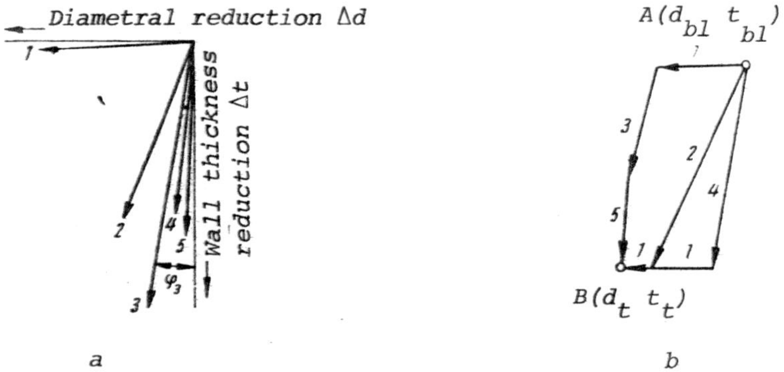

Fig.69

(a) vectors for methods of processing and (b) proportions of their use: 1) drawing without mandrels; 2) drawing with a short secured mandrel; 3) rolling in KhPT mills; 4) drawing with a long mandrel; 5) drawing with KhPTR mills.

For instance, in the case of drawing with no mandrel $\Delta t_{n.m} \approx 0$ and $k_{n.m} \approx 0$; for tube cold rolling mills, $\Delta d_{KhPTR} \approx 0$ and $k_{KhPTR} \approx \infty$ It follows from what has been stated that the direction of the vector $\overline{MN}$ is decided by the method of processing, and its magnitude is governed by the relative extent to which the particular method is used. The approximate direction of the vectors

$$\phi_i = \arctan \frac{\Delta t_i}{\Delta d_i} \tag{155}$$

for the different methods of processing are given in fig.69a

With specific blank and tube dimensions (co-ordinates), transition from point $A(d_{bl}, t_{bl})$ to point $B(d_t, t_t)$ can be made by using one or several methods, as fig.69 shows. Each process is represented by specific consumptions on deformation. To evaluate the consumption C_i it is convenient to use not the absolute reductions $\Delta t_i = t' - t$ and $\Delta d_i = d' - d$, but the relative reductions or coefficients of deformation:

$$\mu_{d_i} = \frac{d'}{d} \text{ and } \mu_{t_i} = \frac{t'}{t} \; *$$

In this case the function $C_i = f(\mu_{d_i}, \mu_{t_i})$ is sufficiently accurately linearized.

We know that the cost of a blank rises as its diameter and wall thickness are reduced. For instance, small volume billets are used for extruding thin walled blanks, and the processing labour (consumption of machine-hours and man-hours per tonne of production) increases. A blank with dimensions represented by point $A_1(d_{bl}, t_{bl})$ will therefore cost more than a blank with dimensions $A_2(d'_{bl}, t'_{bl})$ with $d'_{bl} > d_{bl}$ and $t'_{bl} > t_{bl}$. In other words, with prescribed finished tube dimensions t_t and d_t the cost of the initial blanks drops as the total (overall) coefficients of deformation $\mu_{\Sigma t}$ and $\mu_{\Sigma d}$ for the wall thickness and diameter increase.

Obviously the total outlays on processing will vary for different technologies. The optimum technological schedule is the schedule with which the total outlay is lowest.

Formulation of the problem: For particular processes i $(i = 1, 2, .. n)$ we wish to find the ratio of derivatives μ_{t_i} (or μ_{d_i}) and common factors $\mu_{\Sigma t}$ and $\mu_{\Sigma d}$ such that the outlays, by the optimized criterion for producing tubes with dimensions t_t and d_t will be the lowest possible.

PRODUCTION OF A MATHEMATICAL MODEL

Principal equation for a mathematical model

Let us suppose that the intensity with which any of the methods i is

---

Here the factors μ_{t_i} and μ_{d_i} and the fuller (more complete) deformation oefficients $\mu_{\Sigma t}$ and $\mu_{\Sigma d}$ indicate by how many times the tube wall hicknesses and diameter are reduced. In the problem under consideration it is convenient to use the reduction in internal tube diameter s the basis.

used is represented by the wall thickness and diametral coefficients of deform tion μ_{ti} *and* μ_{di}. The following expressions are then vali for the **overall** (total) and partial deformations:

$$\mu_{\Sigma t} = \frac{t_{bl}}{t_t} = \mu_{t_1} \mu_{t_2} \cdot \cdot \cdot \cdot \mu_{t_n} \tag{156}$$

$$\mu_{\Sigma d} = \frac{d_{bl}}{d_t} = \mu_{d_1} \mu_{d_2} \cdot \cdot \cdot \cdot \mu_{d_n} \tag{157}$$

The minimum blank wall and diameter dimensions $t_{bl(min)}$ *and* $d_{bl(m}$ which lie on the boundary *de* of the regions I and II of the diagram fig. 68 can be specified for any of the methods of producing blanks (extrusion, hot rolling etc.). It is not advisable to produce blank with smaller dimensions for any technological reason (unsatisfactory surface finish, low tool life, force required above the rated force, etc.). The optimum blank dimensions required can be expressed by:

$$t_{bl} = t_{bl\ (min)} \cdot \mu_{tbl} \tag{158}$$

$$d_{bl} = d_{bl\ (min)} \cdot \mu_{dbl} \tag{159}$$

If (158) and (159) are substituted respectively in equations (156) and (157), and the latter reduced to logarithmic form, we obtain the following principal equations for the mathematical form:

$$\sum_{i=1}^{n} log\ \mu_{t_i} - log\ \mu_{t_{bl}} = log\ t_{bl(min)} - log\ t_t \tag{160}$$

$$\sum_{i=1}^{n} log\ \mu_{d_i} - log\ \mu_{d_{bl}} = log\ d_{bl(min)} - log\ d_t \tag{161}$$

<u>Additional equations - limitations</u>

In the processes under consideration, $\mu_{t_i} \geqslant 1$ *and* $\mu_{d_i} \geqslant 1$; theref

$$log\ \mu_{t_i} \geqslant 0 \tag{162}$$

and

$$\log \mu_{d_i} \geqslant 0 \tag{163}$$

Since the ratio of the wall thickness and diameter deformation coefficients is specified for each of the processes:

$$\frac{\mu_{t_i}}{\mu_{d_i}} = k_i^* \tag{164}$$

if equation (164) is reduced to logarithmic form, we get

$$\log \mu_{t_i} - \log \mu_{d_i} = \log k_i^* \tag{165}$$

The limiting parameters (the minimum and maximum tube and blank diameters and wall thicknesses prescribed by the characteristics of the equipment in each *i-th* operation (method) can be represented as follows. If, after the *k-th* operation, the tube diameter and wall thickness must be not less than d_k and t_k

$$\sum_{i=1}^{n} \log \mu_{d_i} - \sum_{i=1}^{k} \log \mu_{d_i} \leqslant \log d_k - \log d_t \quad \text{and} \tag{166}$$

$$\sum_{i=1}^{n} \log \mu_{t_i} - \sum_{i=1}^{k} \log \mu_{t_i} \leqslant \log t_k - \log t_t \tag{167}$$

since

$$\sum_{i=1}^{n} \log \mu_{\theta_i} = \sum_{i=1}^{k} \log \mu_{\theta_i} + \sum_{i=k}^{n} \log \mu_{\theta_i}, \quad \text{and} \tag{168}$$

$$\sum_{i=k}^{n} \log \mu_{\theta_i} = \log \frac{\theta}{\theta_t} \tag{169}$$

where:

θ = the tube parameter (its diameter *d* or wall thickness *t*)

If the maximum permissible blank dimensions for the last $(k + 1)$-th process are specified, after the *k-th* process the diameter and wall thickness of the tube must not exceed a specific magnitude of the tub must not exceed a specific magnitude. This is ensured by altering the sign of the inequalities in expressions (166) and (167).

The purpose function: The minimized criterion can be either the temporary expenditure representing the length of the cycle and the amounts of work (machine-hours and man-hours) or the expenditure in terms of cost (pay, electric power, etc.). Since the diametral and wall thickness deformations for each *i-th* process must satisfy conditions (165),,the expenditure figures for processing the metal can be represented, in the linear expression for the purpose functior in terms of one of the deformation indices.

Let us use C_i to denote the expenditure on a single operation of deformation in the *i-th* cold process, and C_t and C_d to represent the savings on producing blanks with dimensions greater than $d_{bl(min)}$ and $t_{bl(min)}$. The optimum combination of operations will then be expressed by the condition

$$\sum_{i=1}^{n} C_i \log \mu_{\theta_i} - C_t \log \mu_{t_{bl}} - C_d \log \mu_{d_{bl}} = min \qquad (170)$$

If the cost or expenditure of time on producing a blank with the minimum dimensions $C_{tmin} \cdot d_{min}$, is used as the summand of purpose function (170), the sense of the purpose function will be clear. In this case the minimized criterion will be the technological cost or outlay in machine-hours or man-hours on producing a finished tube

The problem can be solved by the simplest method (83), and amounts to finding the values of μ_{ti}, μ_{di}, μ_{dbl} and μ_{dbl} in equations (160-163) and (165-167), with which purpose function (168) will have its minimum value. If it is found, as a result of the solution, tha $\log \mu_{\theta_i} = 0$, i.e. that $\mu_{\theta_i} = 1$, this indicates that the method is not used in the optimum process schedule.

The variation found may be the optimum for producing only particul tubes. Any of our existing shops may produce a wide range of tubes. It is then practically impossible to produce each of the tube dimensions by the optimum technological schedule owing to the quite definite available machine-hours for each type of equipment. In thi case, to obtain the best answer, the optimality of the technological schedules as a whole must be considered (59)

METHOD OF CALCULATING STANDARD TECHNOLOGICAL CHARTS

On the basis of the technological parameters worked out for processes, standard technological charts are compiled for the schedules investigated. The standard technological charts are calculated for a tonne (or 1000m) of finished products, this being the unit by which volume of productions are measured.

The standard technological charts include a list of technological operations, the particulars of the equipment used, the principal dimensions of the blanks after each operation, the yields of sound metal by operations, the norm machine-hours and man-hours, and the estimated and pay figures for the operations per tonne of finished tubes.

The standard process charts must also contain norm consumption figures for electric power, fuel and auxiliary materials. It is particularly important, when standard process charts are calculated that the wastage of metal by operations should be correctly determined, also the yields of sound metal and the consumption coefficients which represent the ratio of the amount of metal reaching an operation to the amount of sound products from the last operation in the process.

Standard process charts calculated to compare different processing techniques include consumption coefficients determined solely from the geometrical wastage in the processes. When the optimum process schedule has been found it is necessary to establish the losses of metal by operations, taking into account the losses by rejection and adjustment of processes carried out by producing balance batches. In order, however, to compare the yield of sound metal with the different technologies, it suffices to calculate the consumption coefficients from the geometrical wastage.

The initial figures for calculating the metal consumption coefficients are the figures for the dimensions and weight of the billets arriving for processing and the blanks produced by extruding the billets, also all the intermediate dimensions of the blanks until a finished tube is produced.

Where there are geometrical lengths of wasted metal, these are determined by finding the difference between the weight of a blank in the previous operation (O_n) and the operation under consideration (O_p) as a percentage of the weight of the billet (O_b)

$$O_i = \frac{O_n - O_p}{O_b} \cdot 100\% \qquad (171)$$

When the wastage and losses are determined as percentages of the specified figures for an operation, this difference is related to the weight of the blank before the operation concerned:

$$O'_i = \frac{O_n - O_p}{O_n} \cdot 100\% \qquad (172)$$

When the percentage wastage is calcualted, the number of blanks obtained from one billet must be taken into account. The total wastage and losses (as %) during the entire technological cycle is determined by totalling the percentage wastages and losses against billet weight for each of the technological operations:

$$O_\Sigma = \sum O_i = \sum \frac{O_n - O_p}{O_b} \cdot 100 \qquad (173)$$

The total yield of sound metal is determined by finding the difference between billet weight, taken as 100%, and the total wastage and losses, found using equation (173):

$$\eta\ (\%) = 100 - O_\Sigma \qquad (174)$$

The consumption coefficients for the mtal throughout the entire technological cycle and for each operation are equal respectively to:

$$L_{r_\Sigma} = \frac{100}{O_\Sigma}\ t/t \qquad (175)$$

$$L_{r_i} = \frac{100}{O_i}\ t/t \qquad (176)$$

If the consumption coefficient is expressed in *kg/t*, the values of L_{r_Σ} and L_{ri} found using equations (175) and (176) are multiplied by 1000.

The expenditures of machine-hours per tonne of sound metal are the norm times for producing unit products multiplied by the consumption coefficient for the operation, since the amount of metal processed to produce a tonne of finished tubes must be greater by the amount of the wastage during the operation concerned.

The consumption of machine-hours (Z) are determined using the equation

$$Z = \frac{1}{H_{v_i}} \cdot \frac{L_{r_i}}{L_{u,e}} \tag{177}$$

Here:

H_{v_i} = the norm production an hour on the *i-th* operation.

The consumption of man-hours Z'_i per tonne of sound metal are calculated in a similar manner:

$$Z'_i = \frac{1}{H_{v_i}} \cdot a_i \cdot \frac{L_r}{L_{u,e}} \tag{178}$$

Here:

a_i = the number of workers on the strength for the *i-th* operation

$L_{u.e}$ = the equipment utilization coefficient by time, allowing for stoppages during shifts.

The estimate per tonne of production is found by multiplying the tariff pay rate for the catergory under consideration by: $\frac{a_i}{H_{v_i}}$

The estimated wages (rates) per tonne of sound metal can be determined by adjusting the product thus obtained by the corresponding consumption coefficient. The time norms, and accordingly the production norms, are established for each operation on the basis of the optimum equipment working conditions considered in the previous chapters. The consumption norms for fuel, electric power, and auxiliary materials per tonne are determined for each operation on the basis of the actual measured figures when balanced batches of tubes are reproduced or of soundly based norms.

The completely calculated standard process charts thus make it possible to compare outputs from the principal production equipment, the work required for the process in machine-hours and man-hours, the metal consumption coefficients and the fuel and electric power consumption norms for each of the technological schedules compared.

Determination of expenditure on production and of the technological cost of a tonne of finished tubes: The consumptions on processing and the direct costs of producing a tonne of finished tubes can be calculated on the basis of indices worked out from the standard technological charts.

Norm consumptions of man-hours per tonne of sound metal (Z_i) calculated using equation (178) are necessary for determining the strength of basic production workers (*Ch*) engaged on producing tubes of a particular size. The number of employees is calculated by the year or by the month. In addition, further additional data are necessary as to the planned production (*P*) of tubes of a particular size, in tonnes a year or a month, and the true available working time per employee (during the same period), ϕ_r.

The basic production manning allowed for by the standard process chart is:

$$CH = \frac{Z'_i \cdot P}{\phi_r} \tag{179}$$

The total number of production workers, i.e. of workers included and not included in the standard process chart, also of auxiliary workers, is determined on the basis of the work force strength *Ch* obtained and the percentage relationship *q* between the categories of workers which have become established in the shop. Table 58 shows the relationship between the categories of workers for one of our existing shops.

The pay figure for all the production workers (B_r) is determined on the basis of the mean monthly production per man, B_{mean}:

$$B_r = B_{mean} \cdot n \frac{Ch}{q} \tag{180}$$

here:

n = the number of months in the period under consideration.

Table 58

Relationship between the Categories of Workers in the tube Shop at One of Our Works.

Serial nos	Categories of workers	proportion q%
1	Production workers included in the SPC*	57.5
2	Production workers not included in the SPC	20.5
3	Auxiliary workers	22.0
	Total	100.0

* SPC is the standard process chart.

The pay for all production workers per tonne of finished tubes (B') is calculated using the equation

$$B' = \frac{B_r}{P} \qquad (181)$$

here: P = the production of tubes of the size under consideration in the period considered. This procedure for calculating the pay of all production workers for a particular technological schedule makes it possible to allow extremely fully for the labour consumed on the production of tubes within a particular range.

It is also recommended that direct calculation should be used for computing the expenditure associated with operating the equipment: depreciation deductions from the cost of equipment, expenditure on electrical power for technological purposes and motors, expenditure on tools and outlay on running repair to equipment and its maintenance.

The total amount of machine-hours required for producing tubes of a particular size is determined on the basis of the consumption z_i of machine-hours for each of the types of equipment per tonne of finished tubes, calculated using equation (177) and the specified production P of finished tubes with the dimensions concerned:

$$Z_{i_{total}} = Z_i \cdot P \qquad (182)$$

On the basis of the actual equipment working time available (by the year or the month), the required number of units of equipment for carrying out the programme, P, is found (or the proportional utilization of the equipment on producing tubes with the particular dimensions concerned):

$$Q_i = \frac{Z_{i_{total}}}{\phi_{u.t}} \qquad (183)$$

Here;

Q_i = the number of units of the *i-th* type of equipment required;
$\phi_{u.t}$ = the true available equipment working time.

The value obtained for Q is the basis employed for determining the cost C_O of the equipment required and the depreciation C_A from its cost:

$$C_0 = C_i \cdot Q_i \tag{184}$$

$$C_A = C_0 \cdot H_{A_i} \tag{185}$$

here:

C_i = the cost of an unit of equipment

H_A = the depreciation norm figure.

The indices Q, Z_{total}, C_0 and C_A are calculated for all the operations in the technological cycle and separately for each type of equipment.

The depreciation deductions from the value of each type of equipment per tonne of finished tubes, C'_{A_i} will be determined by:

$$C'_{A_i} = \frac{C_{A_i}}{P} \tag{186}$$

The expenditure on electric power for carrying out the technology is calulated from the norm consumptions of electric power per tonne of products, taking into account the appropriate consumption coefficients. In casès in which the expenditure on electric power for the technological operations remain unchanged when several technologies are compared, no allowance need be made for it.

The expenditure on electric power for motors is determined on the baisis of details of the installed motor power and the coefficients of utilization of the installed motor power. If the amount of equipment required for producing tubes by a particular technology Q_i is known, having been calculated using equation (183), the consumption of electric power by the motors and the expenditure on electric power per tonne of tubes can be determined using the following equations:

$$Ts_m = \frac{N \cdot \phi \cdot Q_i \cdot \phi_{d.o}}{P} \tag{187}$$

$$Ts_{el} = Ts_m \cdot I \tag{188}$$

here:

Ts_m = the consumption of electric power on the motors per tonne of finished tube.

N – the installed electric motor power in kW
ϕ = the coefficient of utilization of the installed power
Ts_{el} = the expenditure on electric power per tonne of finished tubes;
I = the cost of 1 kWh of power required.

The consumption of electric power by the motors also can be calculated for each type of equipment separately, since the installed motor power and the coefficient of utilization of the installed power are not the same for different types of equipment. The expenditure on tools is determined on the basis of the durability and the cost of a set of tools or a single tool, allowing for the possible number of sharpening or regrinding operations and their cost.

The expenditure on running repairs to equipment is calculated on the basis of details of the repair complexity categories for the mechanical and electrical parts of the equipment. If the work consumed on running repairs to equipment and the annual outlay on repairs are available, the outlay on running repairs to equipment per tonne of finished tubes can be determined using the equation:

$$Z_{r.r_i} = \frac{Ts_i \cdot Q_i}{P} \tag{189}$$

here:

$Z_{r.r_i}$ = the expenditure on running repairs per tonne of finished tubes;
Ts_i = the expenditure on repairing the mechanical and electrical parts of a unit of equipment in a year or the comparison period of time.

The expenditure on the maintenance of equipment is taken as being 1-6% of its cost, for the equipment used with the process schedule concerned. If this expenditure is related to the production of a particular range of tubes it is possible to obtain the expenditure on maintaining equipment per tonne of finished tubes. In many cases, however, because this index is low, it need not be included in practical calculations.

The items of expenditure enumerated, calculated by this procedure embrace up to 75% of all the expenditure on processing. These items vary according to the process schedules compared and the output of the basic production equipment. It must be noted that, to compare process schedules, it is necessary to calculate the economic indices with other conditions maintained constant. This means the calculation for the technologies investigated must be made for the same productions of particular tube dimensions, and that all the indices must be related to a single period of time. The process cost of prodccing a tonne of finished tubes includes the cost of metal, deducting wastage and expenditure on processing.

Calculation of the cost of the equipment required for each technology makes it possible to determine the specific capital outlays for each of the technologies compared, and to compare them in terms of this index. It must be noted that, in this case, the specific capital outlay figures do not by any means allow for all the capital outlay. They do not include the specific capital outlay on transportation equipment, buildings and other structures. It is therefore recommended that different technologies should be compared in terms of the coefficient of capital required, which represents the ratio of the specific capital outlays with the different technologies to one capital outlay taken as the basic outlay.

This system of indices for assessing technical and economic effectiveness when selecting the optimum technology for producing tubes makes it possible to compare different technologies not only as regards metal consumption coefficients, equipment output and work required to produce a tonne of finished tubes, but also as regards expenditure on production. This comparison makes it possible to carry out a thorough evaluation of a particular technological process, to select the most rational process schedule to plan further means of reducing production costs, and to establish limits for the use of a particular process which are rational from the technical and economic points of view. The coefficient of capital outlay on equipment also makes it possible to select the more effective variation, with lower capital outlay, when designing new shops and types of equipment and planning new tehcnologies.

Examples of calculating the principal technical and economic indices for a technological schedule: The procedure for calculating the principal technical and economic indices is illustrated using, as an example, the technology for producing 22 x 20 (1.0) x 7600mm tubes of MNZh5-1 alloy.

In order to analyze the different technologies with an increased number of tubes per blank, experimental batches of 22 x 20 x 7600mm tubes were cold rolled under production conditions and a full calculation of economic effectiveness was made Details of the work required for a number of the production processes, obtained by carrying out balance batch production were used in the calculation. In order to observe the conditions of comparability, the yields of sound metal and consumption coefficients were determined solely, for this process schedule from the geometrical wastage.

Let us now consider calculation of the geometrical wastage and losses for the production of 22.0 x 20.0 x 7600mm tubes of MNZh5-1 alloy by the technological schedule "extrusion - cold rolling", using a KhPT-32-2 tube cold rolling mill (table 59). After filling in the left-hand part of the technological chart and calculating the metal consumption coefficients, the work required on technological operations is calculated. In the case of operations such as extrusion, straightening, pickling and washing blanks, and annealing and cutting up the tubes produced, the true work figures are used, obtained on the basis of measurements made during the technological cycles for these operations, stopwatch recordings and the analysis of the initial data.

When the time norms and process norms were calculated for rolling 22 x 20 x 7600mm tubes of MNZh5-1 alloy, the following technological process parameters wee used; these were obtained from the production of experiments batches of tubes, and also taken from the norm data:

1. According to experimental rolling the linear movement $m\mu_{\Sigma}$ of a tube being rolled was taken as 45mm, the number of passes a minute as $n = 98$

2. The time for entering a fresh tube was taken from stopwatch figures, measured for the KhPT-32-2 mill, as being $t_p = 58s$

3. The time spent on non-cycle operations, including preparatory and concluding times, preventative inspection, and repairs to equipment, and its technical maintenance, also time spent on changing bars, mandrels and rolls, is taken as being 6.7% of shift time (41)

The time norms and processing norms for rolling 22 x 20 (1.0) x 7600mm tubes from 38 x 30 (4.0) x 3300mm blanks are determined in the following sequence. The transverse cross-sectional areas of the blanks and tubes are known: $F_{bl} = 427mm^2$; $F_t = 59.1mm^2$. When the reduction coefficient μ_r for rolling has been determined, the length of rolled tube can be found:

$$\mu_{\Sigma T} = \frac{F_{bl}}{F_t} = \frac{427}{59.1} = 7.23$$

Table 59 Calculating the geometrical wastage and losses and the consumption coefficient for the production of 22.0 x 20.2 x 7600 mm tubes of MNZh5-1 alloy

Technological operations	Dimensions of blank after operation mm			Number of tubes from an ingot	Weight after operation kg		Geometrical wastage on operations				Consumption coefficient per tonne of sound metal kg
								% wastage			
	ℓ	D_{ext}	d_{int}		Blank	Total from ingot	By weight kg (O_n-O_p)	Of prescribed operational wastage	On ingots	Total from ingot before operation	
Heating ingot	310	150	—	1	46,8	—	—	—	—	—	$\frac{1000}{1-0,23}=1300$
Extrusion	10 850	38	30	1	41,2		46,8—41,2= =5,6	$\frac{5,6}{46,8}=12,0$	$\frac{5,6}{46,8}=12,0$	12,00	1145
Cutting blanks	3 300	38	30	3	12,6	37,8	41,2—37,8= =3,4	$\frac{3,4}{41,2}=8,25$	$\frac{3,4}{46,8}=7,26$	19,26	1050
Pickling	3 300	38	30	3	12,6	37,8	—	—	—	—	—
Straightening	3 300	38	30	3	12,6	37,8	—	—	—	—	—
Rolling	23 900	21,8	20	3	12,6	37,8	—	—	—	—	—
Cutting up	7 960	21,8	20	9	4,2	37,8	—	—	—	—	—
Annealing	7 960	21,8	20	9	4,2	37,8	—	—	—	—	—
Cutting to measured lengths	7 600	21,8	20	9	4,0	36,0	37,8—36,0= =1,8	$\frac{1,8}{37,8}=4,76$	$\frac{1,8}{46,8}=3,84$	23,10	1000

$$L_t = \ell_{bl} \cdot \mu_{\Sigma T} = 3300 \cdot 7.23 = 23900mm$$

We took the linear displacement as being $m\mu_{\Sigma} = 45.0$, corresponding to $m = 6.25mm$ with $n = 98$ passes/min. The machine time t_{mach} can then be determined for rolling one blank.

$$t_{mach} = \frac{3300 \cdot 60}{n \cdot m} = \frac{3300 \cdot 60}{6.25 \cdot 98} = 324s$$

The operation time is therefore

$$t_{op} = t_{mach} + t_p = 324 + 58 = 382s$$

The norm production for a 7-h shift will, with $L_{i.o} = 0.937$

$$N_{v_i} = \frac{25200 \cdot 0.937}{382} = 61.5 \text{ items, or } 770kg$$

The coefficient of recharging

$$K_{mach} = 1 - \frac{t_p}{t_{mach} + t_p} = 1 - \frac{58}{382} = 1 - 0.152 = 0.848$$

The hourly output of the KhPT-32 mill when used for rolling 22.0 x 20 (0.9) x 8000mm tubes from 38 x 30 (4.0) x 3300mm blanks is determined using equation (136)

$$M = \frac{7.23 \cdot 6.25 \cdot 98 \cdot 60}{1000} \cdot 0.848 = 225m/h$$

The shift output from the KhPT-32 mill is:

$$M_{sr} = \frac{7.25 \cdot 6.25 \cdot 60}{1000} \quad 98 \cdot 0.848 \cdot 7 \cdot 0.933 = 1470m/shift$$

The time norm per tonne of tubes is

$$\frac{1000 \cdot 7 \ h}{770 \ kg/shift} = 9.09 \ h$$

A standard process chart can be compiled by the procedure set forth in this chapter on the basis of the results obtained (Table 60). Having thus calculated the metal consumption coefficient (Table 59) and the work required to produce a tonne of finished tubes (Table 60) it is next possible to determine the expenditures on processing and the technological cost of the tubes.

Table 60. Standard process chart

Date of production 19 . MNZh5-1 alloy as per GOST. Yields of sound metal: 76.9 from billet (or charge), 95.3% from blanks. Norm. consumption of blank per t of finished products 1300kg. Components "tubes" as per GOST. Final dimensions 22x20x7600mm. Sub-group - Mean mass of 1 finished tube 4.0kg. Billet dimensions dia.150x 310mm blanks dimensions 38x30 (4.0)mm. Mass of 1 blank 46.8kg. Origin of blanks: the foundry shop.

Number of operation	Description of operation	Machinery, number, and working location	Technological particulars of operations	Dimensions after operations mm		
				ℓ	Dxd	t
1	Heating billet	Induction furnace	-	310	150	-
2	Extrusion	1500 t press	-	10 850	38x30	4.0
3	Cutting blanks	disc saw	$\ell_{g.1}$=950mm u=3	3 300	38x30	4.0
4	Pickling	Pickling bath	-	3 300	38x30	4.0
5	Straightening	Straightening machine	-	3 300	38x30	4.0
6	Rolling	KhPT-32 mill	n=98 passes min $m\mu_{\Sigma}$=45mm	23 900	22x20	1.0
7	Cutting in parts	Disc saw	u=3	7 960	22x20	1.0
8	Annealing	Electrical resistance apparatus	-	7 960	22x20	1.0
9	Cutting to measured lengths	Disc saw	$\ell_{g.1}$=360mm	7 600	22x20	1.0
10	Packing	Manual	-	7 600	22x20	1.0
11	Despatch to store	Electric truck	-	7 600	22x20	1.0

* The first figures is the number of workers, the second their category. In column 1, $\ell_{g.1}$ stands for "length along generating line"

Weight of 1 item after operation	Norm.wastage and losses: Totals for operations: % of specified for operations	Norm.wastage and losses: Totals for operations: with relation to initial operation	Norm.wastage and losses: Totals as % from start of previous operation	Correction factor kg per kg of finished metal	Norm.utilisation of equipment in times as coefficient	Norm.strengths of workers on operations by categories	Norm.production per shift	Norm.consumption of machine hours per t of finished product	Norm.consumption of man-hours per t of finished product
46.8	-	-	-	1300	-	-		-	
41.2	12.0	12.0	12.0	1245	0.92	1-6* 2-5 4-4	13.2	0.649	4.606
12.6	8.25	7.26	19.26	1050	-	-			
12.6	-	-	-	1050	0.97	1-3	15.0	0.488	0.488
12.6	-	-	-	1050	-	1-3	34.2	0.215	0.215
12.6	-	-	-	1050	0.9	1-6	0.71	10.35	10.35
4.2	-	-	-	1050	-	-	-	-	-
4.2	-	-	-	1050	0.95	2-4	13.5	0.546	1.092
4.0	4.76	3.84	23.10	1000	-	2-3	28.0	0.263	0.526
4.0	-	-	-	1000	-	1-3	6.0	-	1.16
4.0	-	-	-	1000	-	2-4	25.0	-	0.56
								12.251	18.997

Table 61

Calculations of Numbers of Production Workers for Rolling MNZh5-1 Alloy Tubes with dimensions of 22 x 20 (1.0) x 7600mm

Operations	Man-hours per tonne of finished	Annual production t	Man-hours on annual production	Available working time per man	Number of men
Extrusion	4.606	421.2	1945.0	1848	1.05
Pickling	0.488	421.2	205.55	1848	0.111
Straightening	0.215	421.2	90.13	1848	0.05
Rolling	10.35	421.2	4360.0	1848	2.36
Annealing	1.092	421.2	459.11	1848	0.24
Cutting to measured length	0.526	421.2	221.55	1848	0.12
Packing in boxes	1.160	421.2	488.59	1848	0.26
Despatch to store	0.56	421.2	235.87	1848	0.13
Total					4.321

Table 62

Calculation of the mean list working strength and pay per tonne of 22 x 20 (1.0) x 7600mm finished tubes of MNZh5-1 Alloy

Description	List strength	Proportion %
Production workers allowed for in NTK	4.321	57.5
Production workers not allowed for in NTK	1.538	20.5
Auxiliary workers	1.645	22.0
Total workers	7.504	100.0

The process figure is first calculated. Since the standard technological chart does not take into account the pay for all the production workers, and only allows for pay according to the tariff scale, we first determine the total number of production workers.

The total strength of production workers (table 61) is calculated in the following manner:

1. The annual consumption of man-hours on production is determined on the basis of the figures, in the standard process chart, for the man-hours consumed on each of the operations and the annual production of finished 22 x 20 x 7600mm tubes in the previous year, the amount being 421.2t

2. The available working time per man is 1848 h a year. These figures make it possible to determine the work force from the standard process chart for the annual production.

The total working strength over the year is determined by means of the steady relationship between groups of production workers (Table 62) Knowing the true level of mean monthly pay to a man (116.2 roubles) we find the pay for all the production workers for a year (1045.12 roubles), and we then determine the wages per tonne of finished tubes, which comes to 24.80 roubles.

The number of machine-hours required for producing the finished tubes in the year is calculated on the basis of the machine-hours consumed on the individual operations and the annual production of finished tubes. On the basis of the actual available equipment working time in a year, the extent to which the equipment is utilized in production tubes of particular dimensions is determined. This quantity is the basis for determining depreciations, expenditure on electric power for motors, running repairs to equipment and its maintenanace and operations.

According to works figures, the coefficient of utilization of installed electric motor power is 0.85 for the rolling mill equipment, and it is 0.65 for the other equipment. The depreciation deductions are calculated on the basis of the balance cost of the equipment in the shop. The expenditure on maintaining and operating equipment is taken as 1% of the cost of the equipment required for producing particular products, as was pointed out earlier. All the figures listed above are given in Table 63.

After deducting wastage the cost of the metal is determined from the work figures for the cost of 1 ton of billets (778 roubles) and

Table 63

Calculation of amount of equipment depreciation deductions, expenditures on electric power, running repairs to equipment, maintenance and operation of equipment, for producing 22x20x7600mm tubes of MNZh5-1 alloy (annual production 421.2t)

Name of equipment	Machine-hours per tonne of finished tubes	Machine-hours on annual production	True available equipment working time h	Equipment required	Cost of equipment roubles		Depreciation deductions	
					unit	amount needed	%	amount roubles
1500t press	0,659	275	5723	0,048	73 900—00	3547—20	8,6	305—05
Pickling baths	0,488	205,65	3429	0,060	6 096—00	365—76	22,8	83—39
Straightening machine	0,215	90,13	5180	0,017	5 657—70	98—44	12,1	11—91
KhPT-32 2 mill	10,35	4350	3888	1,12	170 289—48	191 000	11,0	21 000
"Snid" apparatus	0,546	229,55	4527	0,05	9 941—80	497—00	15,0	74,56
1500t press	0,263	235,87	5738	0,04	125—00	5—00	10,9	0,545
TOTALS								
by programme	—	—	—	—	—	195513-49	—	21475-76
by tonnes of finished tubes	—	—	—	—	—	465—0	—	51—00

1 ton of wastage (700 roubles for the MNZh5-1 alloy): *778·1300 - 700·0.300 = 800 roubles*. The expenditure on processing obtained by calculation, with the costs of 1 ton of finished tubes, are thus, (in roubles/kopecks)

Cost of metal after deducting wastage	800-00
Expenditure on processing:	
pay	24-80
depreciation	51-00
electric power	3-94
running repairs	15-96
maintenance	4-65
Total expenditure on processing	100-35
Total technological cost	900-35

Installed motor power kW		Electric power consumption kWh/year	Cost of annual power consumption roubles	Outlays on running repairs roubles				Expenditure on maintenance and operation of equipment roubles
				Mechanical part		Electric part		
unit	amount in use			unit	number in use	unit	number in use	
45,65 —	2,19 —	12 580 —	234—87 —	4820-34 —	231—87 —	216—12 —	10—80 —	35—47 3—65
5,2	1,12	2 490	4—49	131—53	2—28	44-99	0—77	0—98
244,5 — 2,8	273 — 0,112	904 000 — 1 270	16 300 — 7,52	5328—65 — 66,08	5960 — 2—64	405—60 65—78 21,60	515 3,28 0,86	1910 4,97 0,05
— —	— —	— —	1658—89 3—94	— —	6196—69 14—70	— —	530-71 1—26	1955—12 4—65

SETTING UP A MATHEMATICAL MODEL FOR CALCULATING THE OPTIMUM TECHNOLOGICAL SCHEDULES FOR TUBE PRODUCTION WITH ELECTRONIC COMPUTERS

The method of setting up mathematical models which can be used in general is the linear programming method, this consists of compiling and solving a system of linear equations in order that the particular solution shall provide a definite (minimum or maximum) value of the purpose function (58,94)

The process of preparing and solving problems of this type with mathematical machinery can conditionally be subdivided into four stages (100):

Stage 1 - working out the mathematical model and determing the optimality criterion. In this process it is necessary to determine the real links, establish the functional relationships, and express them in the form of mathematical equations, i.e. to formalize the problem

Stage 2 - preparation of the problem for solution with an electronic computer; first the required information is collected and the initial data formulated, and secondly the mathematical apparatus is worked out - the algorithm and work programme.

Stage 3 - calculation with the electronic computer.

Stage 4 - analysis of the results obtained, since extreme problems are ususally solved in terms of several optimality criteria.

Mathematical formulation of the problem: It is required that various standard tube dimensions $j (j=1. 2. ...., n)$ with an amount Q_j of each of the dimensions, shall be produced. Various different technologies i $(i=1, 2, ..., m)$ can be used, with various types of equipment - r *plants* $(r=1, 2, ..., k)$.

The intensity of utilization of a particular technology i for producing type j production is governed by the amount of production x_{ji} produced with the technologies concerned. We know that the outlay on producing unit of product j by technology i is governed by the index C_{ji}. The total must satisfy the condition:

$$\sum_{j=1}^{n} \sum_{i=1}^{m} C_{ji} x_{ji} = C\,(min) \qquad (190)$$

If the task of determining the combination of technological schedules to produce the maximum profit is solved (the profit is the difference between the wholesale price S_j of a product and its cost s_{ji})

$$\sum_{j=1}^{n} \sum_{i=1}^{m} (S_j - s_{ji})\, x_{ji} = C\,(max) \qquad (190a)$$

Conditions (190) and (190a) are purpose functions, and the multiples C_{ji} and $(S_j - s_{ji})$ are the criteria optimized.

In addition to the purpose function, the mathematical model of the optimum process schedules (rational utilization of equipment) must include additional limitation conditions (83).

a) the total volume of production for each standard tube dimension must correspond to the plan target Q_j:

$$\sum_{i=1}^{m} x_{ji} = Q_j \tag{191}$$

b) the variables x_{ji} must be varied within the limits

$$x_{ji} \leqslant 0 \tag{192}$$

since negative values have no actual significance;

c) when unit products of type j are produced by variation i in the equipment (or plant) of type r, t_{jir} machine-hours are consumed to carry out the plan j, it is therefore necessary that

$$\sum_{i=1}^{m} t_{jir} x_{jr} \leqslant T_r \tag{193}$$

where:

T_r = the available machine-hours for plant r for the planned period of time

The problem amounts to finding values of x_{ji} with which purpose functions (190) or (190a) will have their minimum (or maximum) values. The case $x_{ji} = 0$ indicates that the technology concerned applied to the mathematical model planned programme, is not rational. When the matrix form of the problem is considered it obviously relates to the distribution category, and can be solved by the method of successive improvement of the plan (the simplex method) an algorithm for which has been worked out and has been extensively used (58, 93, etc).

The procedure for determining the optimum process schedules and the rational utilization of the equipment would be selected for the example of the production of 19 x 1.5, 40 x 2.5 and 120 x 5mm copper tubes using the following equipment: 1) a 3000t press 2) a 1500t press; 3) a KhPT-55 cold rolling mill; 4) a KhPT-32 cold rolling mill; 5) a 30t draw bench; 6) a 10t draw bench; 7) a 5t draw bench

In the example we are considering the monthly production of tubes with the equipment mentioned must be as follows: a)19x1.5-80t; b) 40 x 2 - 100t; c) 120 x 5mm - 70t. These tubes can be produced by the following technological schedules.

Dimensions 19 x 1.5mm:

1. Extrusion of 34 x 2mm blanks in a 1500t press from 150 x 250mm billets and drawing the blanks in three passes down to the final dimensions in a 5t draw bench.

2. Extrusion of 36 x 5mm blanks from 175 x 250mm billets in a 1500t press and cold rolling the blanks down to the final dimensions in a KhPT-32 mill.

3. Extrusion of 40 x 7mm blanks in a 1500t press from 175 x 250mm billets, cold rolling the billets down to 20.6 x 1.8mm in a KhPT-32 mill, and drawing in a single pass down to the final dimensions in a 5t draw bench.

4. Extrusion of 63 x 11.5mm blanks in a 3000t press from 200 x 375mm billets, cold rolling the blanks down to 36 x 5mm in a KhPT-55 mill, and cold rolling down to the final dimensions in a KhPT-32 mill.

Dimensions 40 x 2.5mm

5. Extrusion of 52 x 3.5mm blanks in a 1500t press from 175 x 350mm billets and drawing the blanks down to the final dimensions in two passes in a 10t draw bench

6. Extrusion of 63 x 11.5mm blanks in a 3000t press from 200 x 375mm billets and cold rolling the blanks down to the final dimensions in a KhPT-55 mill.

7. Extrusion of 63 x 11.5mm blanks in a 3000t press from 200 x 375m billets, cold rolling the blanks down to 43 x 3mm in a KHPT-55 mill, and drawing down to the final dimensions in a single pass in a 10t draw bench.

Dimensions 120 x 5mm;

8. Extrusion of 133 x 6.5mm blanks in a 3000t press from 300 x 475mm billets, drawing down to the final dimensions in two passes in a 30t draw bench.

9. Extrusion of 173 x 6.5mm blanks in a 3000t press from 400 x 550mm billets, and drawing down to the final dimensions in three passes in a 30t draw bench.

No doubt there are many other different technological scheduled for producing tubes of the dimensions indicated with the available equipment, but to consider all the possible technologies is not the purpose of this work.

Table 64 contains details of the machine-hours consumed on the production of 1 ton of finished tubes by each of the process schedules, taking into account the consumption of metal (the correction coefficients) on each operation* The last column of Table 64 gives the profit - the difference between the wholesale price and the total cost of producing 1 ton of tubes by each of the process schedules. The bottom line gives the available machine-hours for each of the types of equipment, these being difference between the calendar time and the time spent on planned preventive repairs and other planned stoppages.

Let us consider the problem of determining the optimum technological schedules and rational utilization of equipment to produce the maximum profit and fulfil the production programme for particular ranges of tubes and volumes of production. The problem is solved by the simple method (58, 83) An equation is compiled for the utilization of each type of equipment for the month taking into account all the technological schedules in which the particular type of equipment is used. The general form of the equation for the utilization of equipment was given earlier.

For instance, the utilization of a 3000t press is determined by means of the equation :

$$0.59x_4 + 0.59x_6 + 0.64x_7 + 0.68x_8 + 0.58x_9 \leqslant 500 \qquad \text{(a)}$$

(see Table 64) For the example under consideration, seven equations are compiled similarly for the utilization of equipment (according to the number of types of equipment). In addition, a further three equations are compiled for determining the fulfilment of the production programme for the prescribed range of tubes and volumes of production.

$$x_1 + x_2 + x_3 + x_4 = 80; \quad x_5 + x_6 + x_7 = 100; \quad x_8 + x_9 = 70 \text{(b)}$$

This system of equation is a mathematical model of the production problem under consideration.

In the example concerned (obtaining the maximum profit while fulfilling the production programme) the optimum criteria (190 and (190a) take the following mathematical form (linear form.

**The details of the machine-hours consumed are taken from the standard technological charts calculated for each of the technology variations*

Table 64

Consumption of Machine hours and Profits

Tube dimensions mm	Schedule no	Machine hours consumed on producing 1 ton of finished tubes with the equipment below *							Profit per tonne of finished tubes ** roubles
		3000t press	1500t press	KhPT-55 mill	KhPT-32 mill	Draw benches 30t	Draw benches 10t	Draw benches 5t	
19x1.5	1	0	0.89	0	0	0	0	4.7	325
	2	0	0.66	0	8.15	0	0	0	335
	3	0	0.68	0	9.15	0	0	2.2	322
	4	0.59	0	9.4	6.12	0	0	0	320
40x2.5	5	0	0.86	0	0	0	3.8	0	170
	6	0.59	0	7.4	0	0	0	0	180
	7	0.64	0	6.9	0	0	2.2	0	165
120x5	8	0.68	0	0	0	2.5	0	0	200
	9	0.58	0	0	0	3.8	0	0	210
		Monthly available machine-hours							
		500	500	380	380	400	400	400	

* The figure 0 indicates that the products concerned are not processed in this equipment with the technological schedule mentioned

** The figures are nominal

purpose function (see table 64):

$$325x_1 + 335x_2 + 322x_3 + 320x_4 + 170x_5 + 180x_6 + 165x_7 + 200x_8 + 210x_9 = C\ (max) \quad (c)$$

In order to solve the system of equations which have been obtained by the simple method, the seven inequalities types (a) are transformed into an equation by introducing additional unknowns x_{10} - x_{16} into them; these equations denote the remaining available machine hours for the equipment after fulfilling the production programme. For instance, the monthly utilization of the 3000t press will now be determined by the equation:

$$0.59x_4 + 0.59x_6 + 0.64x_7 + 0.68x_8 + 0.58x_9 + x_{10} = 500$$

In addition to this, in order to find the initial key plan (solution) for the systems of equations obtained, and to obtain a single matrix in the initial simplex table, additional (fictitious) unknown x_{17}-x_{19} are introduced into equation (b).. For instance the first equation in system (b) will now take the form

$$x_1 + x_2 + x_3 + x_4 + x_{17} = 80$$

where x_{17} is the possible under-fulfilment of the programme for producing 19 x 1.5mm tube when insufficient machine-hours are available for any particular type of equipment or for other reasons. It should be noted that the additional unknowns x_{10}-x_{19} are not all included in linear equation (c) (the profit obtained from the production of 1 ton of tubes by each technological schedule (S_j-S_{ji}) is equal to 0 with these unknowns) since, obviously, under-fulfilment of the production programme and unused equipment machine-hours do not provide a profit. Moreover it is necessary that $x_j \geqslant 0$, since negative values of x_j have no real meaning.

The system of equations can be written in vector form:

$$\bar{P}_1x_1 + \bar{P}_2x_2 + \ldots + \bar{P}_{19}x_{19} = \bar{P}_0$$

here:

$\bar{P}_1$ = the vector containing all the x_1
$\bar{P}_2$ = the vector containing all the x_2 etc.,
$\bar{P}_0$ = the vector containing all the free terms.

All the conditions and requirements for the problem can conveniently be tabulated in the initial simplex table (Table 65).

Table 65 Initial simplex table

Basic vectors	c_{ij}	$\bar{p}_1$	$\bar{p}_2$	$\bar{p}_3$	$\bar{p}_4$	$\bar{p}_5$	$\bar{p}_6$	$\bar{p}_7$	$\bar{p}_8$	$\bar{p}_9$	$\bar{p}_{10}$	$\bar{p}_{11}$	$\bar{p}_{12}$	$\bar{p}_{13}$	$\bar{p}_{14}$	$\bar{p}_{15}$	$\bar{p}_{16}$	$\bar{p}_{17}$	$\bar{p}_{18}$	$\bar{p}_{19}$	$\bar{p}_0$
		325	335	322	320	170	180	165	200	210	0	0	0	0	0	0	0	0	0	0	
$\bar{p}_{10}$	0	0	0	0	0,59	0	0,59	0,64	0,68	0,58	1	0	0	0	0	0	0	0	0	0	500
$\bar{p}_{11}$	0	0,89	0,66	0,68	0	0,86	0	0	0	0	0	1	0	0	0	0	0	0	0	0	500
$\bar{p}_{12}$	0	0	0	0	9,4	0	7,4	6,9	0	0	0	0	1	0	0	0	0	0	0	0	380
$\bar{p}_{13}$	0	0	8,15	9,15	6,12	0	0	0	0	0	0	0	0	1	0	0	0	0	0	0	380
$\bar{p}_{14}$	0	0	0	0	0	0	0	0	2,5	3,8	0	0	0	0	1	0	0	0	0	0	400
$\bar{p}_{15}$	0	0	0	0	0	3.8	0	2.2	0	0	0	0	0	0	0	1	0	0	0	0	400
$\bar{p}_{16}$	0	4,7	0	2,2	0	0	0	0	0	0	0	0	0	0	0	0	1	0	0	0	400
$\bar{p}_{17}$	0	1	1	1	1	0	0	0	0	0	0	0	0	0	0	0	0	1	0	0	80
$\bar{p}_{18}$	0	0	0	0	0	1	1	1	0	0	0	0	0	0	0	0	0	0	1	0	100
$\bar{p}_{19}$	0	0	0	0	0	0	0	0	1	1	0	0	0	0	0	0	0	0	0	1	70
z_j		0	0	0	0	0	0	0	0	0	0	0	0	0	0	0	0	0	0	0	0
$C_j - z_j$		325	335	322	320	170	180	165	200	210	0	0	0	0	0	0	0	0	0	0	

In the initial solution (the reference plan). the profit was equal to zero (the column $\bar{P}_0$, *the line for* x_j). To obtain a profit, it is necessary to include one of the types of tube in production, i.e. using mathematical terminology, to go over to a new basis, replacing one of the vectors $\bar{P}_1 - \bar{P}_9$, The conversion from basis to basis is made in a single full elimination step. This is done by the procedure given in reference (58 and 83).

Tubes should be produced by the technological schedule which provides the maximum profit, i.e. it is necessary to determine the vector introduced into the basis (to find the key column). For this purpose the value of the function $(C_j - z_j)$ signifying the possible increase in the profit when tubes are produced by each technological schedule with the variation of utilisation of equipment concerned is determined for each column. The values z_j are determined using the equation:

$$z_j = \sum_{1}^{i} C_i t_{jir}; \qquad (194)$$

here:

C_i = the basic vector coefficient in linear form;
t_{jir} = the consumption of machine-hours
z_j = the profit obtained when the tubes are produced by each technological schedule with the variation of equipment utilization concerned, and equal to $S_j - S_{ji}$

For instance, the value of $(C_j - z_j)$ *for the* vector $\bar{P}_1$ will be:

$$(C - z) = 325 - (0 \cdot 0 + 0 \cdot 0.89 + 0 \cdot 0 + 0 \cdot 0 + 0 \cdot 0 + 0 \cdot 0 + 0 \cdot 4.7 + 0 \cdot 1 + 0 \cdot 0 + 0 \cdot 0) = 325$$

The values of $(C_j - z_j)$ are determined in an equivalent manner for all the other vectors. The vector which must be introduced into the basis is the vector with the maximum value of $(C_j - z_j)$. In our example, this vector is $\bar{P}_2$ (see Table 65). Tubes with dimensions of 19 x 1.5mm must thus be produced by the 2nd technological schedule which provides the maximum profit.

The maximum production of 19 x 1.5mm tubes by technology 2 is limited by the full "bottleneck" load, which is governed by the minimum value of the ratio Q_i/t_i for each type of equipment taking part in the production of tubes by the technlogical schedule concerned.

In the example we are considering, the "bottleneck" is the KhPT 32 mill, since its ratio

$$\frac{Q_i}{t_i} = \frac{380}{8.15}$$ is lower than the ratio $$\frac{Q_i}{t_i} = \frac{500}{0.66}$$ for the 1500t press

The vector taken from the basis is thus determined (the key line is found) In our case this line is the line for $\bar{P}_{13}$

The key element the value of which is necessary for plotting the next simplex table (Table 66) is found at the intersection of the key column and the key line. In the case we are considering, the key element is the number 8.15.

The vector $\bar{P}_2$ introduced into the basis of the next simplex table must be unitary, i.e. its key element must be equal to unity and the other coefficients to zero. For this purpose all the digits in key line $\bar{P}_{13}$ are divided by the key number (8.15) This is how the line for the $\bar{P}_2$ in the new simplex table is obtained.

The other numbers in the new table are obtained in the following manner. Those lines in the initial table which have coefficients 0 in the key column are transferred unchanged to the new table. In our case these are the lines for $\bar{P}_{10}$; $\bar{P}_{12}$; $\bar{P}_{14}$; $\bar{P}_{15}$; $\bar{P}_{16}$; $\bar{P}_{18}$; $\bar{P}_{19}$ The lines for $\bar{P}_{11}$ and $\bar{P}_{17}$ have coefficients differing from zero in the key column. The line $\bar{P}_2$ in the new table is multiplied by these coefficients, and the products are subtracted from the corresponding coefficients for these lines ($\bar{P}_{11}$ and $\bar{P}_{17}$) in the initial table The calculated differences are the coefficients for the lines for $\bar{P}_{11}$ and $\bar{P}_{17}$ in the new table. For instance, the coefficient in column P_o, the line for $\bar{P}_{11}$, in the new table becomes:

$$500 - 0.66 \cdot \frac{380}{8.15} = 469.2$$

This figure denotes the remaining machine hours for the 1500t press after producing the blanks for 19 x 1.5mm tubes by technological schedule 2.

Table 66 also shows that the available machine-hours for the KhPT-32 mill were fully utilized (the vector $\bar{P}_{13}$ is not included in the basis). The amount of 19 x 1.5mm tubes produced, out of 80t, was 46.6t (column $\bar{P}_o$ and line $\bar{P}_2$), and the profit came to 15 600 nominal units (column P_o and line z_j). To increase these profits it is necessary to conclude the programme for producing 19 x 1.5mm tubes using

Table 66 Simplex table (improved plan)

Basic vectors	C_{ij}	$\bar{p}_1$ 325	$\bar{p}_2$ 335	$\bar{p}_3$ 322	$\bar{p}_4$ 320	$\bar{p}_5$ 170	$\bar{p}_6$ 180	$\bar{p}_7$ 165	$\bar{p}_8$ 200	$\bar{p}_9$ 210	$\bar{p}_{10}$ 0	$\bar{p}_{11}$ 0	$\bar{p}_{12}$ 0	$\bar{p}_{13}$ 0	$\bar{p}_{14}$ 0	$\bar{p}_{15}$ 0	$\bar{p}_{16}$ 0	$\bar{p}_{17}$ 0	$\bar{p}_{18}$ 0	$\bar{p}_{19}$ 0	$\bar{p}_0$
$\bar{p}_{10}$	0	0	0	0	0,59	0	0,59	0,64	0,68	0,58	1	0	0	0	0	0	0	0	0	0	500
$\bar{p}_{11}$	0	0,89	0	—0,08	—0,05	0,86	0	0	0	0	0	1	0	—0,08	0	0	0	0	0	0	469,2
$\bar{p}_{12}$	0	0	0	0	9,4	0	7,4	6,9	0	0	0	0	1	0	0	0	0	0	0	0	380
$\bar{p}_2$	335	0	1	1,12	0,75	0	0	0	0	0	0	0	0	0,12	0	0	0	0	0	0	46,6
$\bar{p}_{14}$	0	0	0	0	0	0	0	0	2,5	3,8	0	0	0	0	1	0	0	0	0	0	400
$\bar{p}_{15}$	0	0	0	0	0	3,8	0	2,2	0	0	0	0	0	0	0	1	0	0	0	0	400
$\bar{p}_{16}$	0	4,7	0	2,2	0	0	0	0	0	0	0	0	0	0	0	0	1	0	0	0	400
$\bar{p}_{17}$	0	1	0	—0,12	—0,25	0	0	0	0	0	0	0	0	—0,12	0	0	0	1	0	0	33,4
$\bar{p}_{18}$	0	0	0	0	0	1	1	1	0	0	0	0	0	0	0	0	0	0	1	0	100
$\bar{p}_{19}$	0	0	0	0	0	0	0	0	1	1	0	0	0	0	0	0	0	0	0	1	70
z_j		0	335	375	250	0	0	0	0	0	0	0	0	40,2	0	0	0	0	0	0	15 600
C_j-z_j		325	0	—53	70	170	180	165	200	210	0	0	0	—40,2	0	0	0	0	0	0	

process schedule no.1 (the maximum value of $C_j - z_j = 325$ relates to the vector P_1). The vector P_1 must be introduced into the basis in place of the vector P_{17} for which the ratio :

$$\frac{Q_i}{t_i} = 33.4 \text{ is the minimum etc.}$$

The rational process schedules are selected in an equivalent manner, and the utilization of equipment for carrying out a production programme for producing 40 x 2.5 and 120 x 5mm tubes is calculated. The calculations end when all possibility of increasing the profit has been exhausted, i.e. when all the coefficients in the line for $C_j - z_j$ are $\leqslant 0$. The final optimum plan, for the example under consideration is given in Table 67.

It follows from Table 67 that, to obtain the maximum possible profit under the particular conditions concerned (58750 nominal units, column $\bar{P}_o$ and line z_j) it is necessary to produce 33.4t of the 19 x 1.5mm tubes by technology 1 (column $\bar{P}_o$ and line P_1) and 46.6t by technology 2 (column $\bar{P}_o$ and line P_2). Since $x_1 = 33.4t$, $x_2 = 46.6t$, $x_3 = 0$ and $x_4 = 0$, $x_1 + x_2 + x_3 + x_4 = 80t$. In the conditions we are considering, it would not be rational to produce the tubes concerned by technologies 3 or 4.

The 40 x 2.5mm tubes must be produced by technologies 5 and 6; technology 7 is not rational ($x_5 = 48.7t$; $x_6 = 51.3t$; $x_7 = 0$; $x_5 + x_6 + x_7 = 100t$).

The 120 x 5mm tubes must only be produced by technology 9; technology 8 is not rational ($x_8 = 0$; $x_9 = 70$; $x_8 + x_9 = 70$)

When the production programme had been carried out, the following equipment machine-hours remained unutilized:

3000t press	429.1 ($x_{10} = 429.1$)
1500t press	397.4 ($x_{11} = 397.4$)
30t draw bench	134 ($x_{14} = 134$)
10t draw bench	215 ($x_{15} = 215$)
5t draw bench	243 ($x_{16} = 243$)

The available machine-hours for the KhPT-55 mill and KhPT-32 mill were fully utilized ($x_{12} = 0$; $x_{13} = 0$). The production programme was fully completed ($x_{17} - x_{19} = 0$). Obviously when the conditions of the problem are altered or the same problem solved in terms of another optimality criterion, there will be a different particular solution.

Table 67 Simplex table (optimum plan)

Basic vectors	C_{ij}	$\bar{p}_1$	$\bar{p}_2$	$\bar{p}_3$	$\bar{p}_4$	$\bar{p}_5$	$\bar{p}_6$	$\bar{p}_7$	$\bar{p}_8$	$\bar{p}_9$	$\bar{p}_{10}$	$\bar{p}_{11}$	$\bar{p}_{12}$	$\bar{p}_{13}$	$\bar{p}_{14}$	$\bar{p}_{15}$	$\bar{p}_{16}$	$\bar{p}_{17}$	$\bar{p}_{18}$	$\bar{p}_{19}$	$\bar{p}_0$
		325	335	322	320	170	180	155	200	210	0	0	0	0	0	0	0	0	0	0	
$\bar{p}_{10}$	0	0	0	0	−0,16	0	0	0,09	0,1	0	1	0	−0,08	0	0	0	0	0	0	−0,58	429,1
$\bar{p}_{11}$	0	0	0	0,03	0,82	0	0	−0,06	0	0	0	1	0,116	0,03	0	0	0	−0,89	−0,86	0	397,4
$\bar{p}_6$	180	0	0	0	1,27	0	1	0,93	0	0	0	0	0,135	0	0	0	0	0	0	0	51,3
$\bar{p}_2$	335	0	1	1,12	0,75	0	0	0	0	0	0	0	0	0,12	0	0	0	0	0	0	46,6
$\bar{p}_{14}$	0	0	0	0	0	0	0	0	−1,3	0	0	0	0	0	1	0	0	0	0	−3,8	134,0
$\bar{p}_{15}$	0	0	0	0	4,83	0	0	1,93	0	0	0	0	0,51	0	0	1	0	0	−3,8	0	215,0
$\bar{p}_{16}$	0	0	0	2,76	−1,18	0	0	0	0	0	0	0	0	0,56	0	0	1	−4,7	0	0	243,0
$\bar{p}_1$	325	1	0	−0,12	0,25	0	0	0	0	0	0	0	0	−0,12	0	0	0	1	0	0	33,4
$\bar{p}_5$	170	0	0	0	−1,27	1	0	0,07	0	0	0	0	−0,135	0	0	0	0	0	1	0	48,7
$\bar{p}_9$	210	0	0	0	0	0	0	0	1	1	0	0	0	0	0	0	0	0	0	1	70,0
z_j		325	335	336	344	170	180	210	210	210	0	0	1,35	1,2	0	0	0	325	170	210	
C_j-z_j		0	0	−14	−24	0	0	−15	−10	0	0	0	−1,35	−1,2	0	0	0	−325	−170	−210	58750

It should be noted that, to a certain extent, the problem considered is conditional. The range of tubes produced in tube shops at present operating is far wider as regards alloys and dimensions, and the available equipment is also more extensive. The mathematical model of the production problem therefore also proves to be far more complex, and solving it manually may prove too laborious or impossible in practice. In this case it is extremely effective to use electronic computers with a sufficient operative storage device capacity. There is no particular difficulty in working out programmes for solving problems of this type with an electronic computer.

It is important to emphasize that it is impossible to use mathematical methods for selecting rational process schedules for producing tubes and for attaining the optimum utilization of equipment without first working out variations of the technology on a technical basis, and making a technical and economic evaluation of these, on a scientific basis, from the point of view of the consumption of machine-hours and man-hours per tonne of finished tubes, together with adjustment coefficients, the cost of processing in each operation, the cost of metal, the process (shop) and total costs, etc.

SCHEDULING

The problem of scheduling is very important in organizing production process; the problem is to establish the most rational distribution of different operations over a period of time. For instance it is extremely important, in the case of shops carrying out extrusion of metals, to know the sequence in which the tube blanks and finished tubes must be produced in the first equipment (the presses) for the last equipment (the KhPT mills and the draw benches) to work without stoppages due to the non-availability of blanks, The solution to this problem will make it possible to have the minimum amount of uncompleted production (semi-finished products) in a shop, and to utilize floor space as well as possible.

The problem of scheduling should be solved after the process schedules for shops under consideration have been determined. Linear programming methods (83, 93) are also used during this. A.L.Lur'e's mathematical model, given in reference (83), is described below, interpreted for tube production.

Mathematical formulation of the problem: From the production plan we know the types of tube $j = 1, 2, \ldots n$, also the amount Ω_j of them which must be produced.

The operations (types of equipment $r = 1, 2, \ldots\ldots k$) and their sequence in time are known for each type of product, from the optimum process shcedules. Since the sequence of operations may vary for different types of tube, the operations are designated by the double indices ij, where j is the type of product and i the serial number of the operation for a type j product. For each type j of tube, it is assumed that $i = 1, 2, \ldots, m_j$, *where* m_j is the number of the last (final) operation of those to which tubes with the number j must be subjected (there may be different numbers of operations for different types of tubes and technological schedules).

The purpose function. It is necessary to find a calendar chart for the utilization of equipment such that the period T during which all the operations for producing the prescribed amount of products are fulfilled is the shortest:

$$T = min \tag{195}$$

The limitation conditions. The total time consumed on processing tubes of type j in equipment r between the initial moment of time $t = 0$ and a moment of time $t > 0$ is a steadily rising function of time $F_{rj}(t)$. At the initial moment of time ($t = 0$), the function $(F_{rj})(t)$ is zero: $F_{rj}(0) = 0$. With $t_2 > t_1$ the following conditi... must be fulfilled:

$$F_{rj}(t_2) - F_{rj}(t_1) \geqslant 0. \tag{196}$$

In each of the plants r, different types of products j are being processed at the same moment of time t i.e. simultaneously. It follows from this that if we total the time taken to produce differ... types of tubes j in the same equipment r for some period between t_1 *and* t_2. This condition is written as follows: with $t_2 > t_1$

$$\sum_{j=1}^{n} \left[F_{rj}(t_2) - F_{rj}(t_1)\right] \leqslant t_2 - t_1 \tag{197}$$

During the plan fulfillment period T, when an amount Q_j of type j tubes is produced all the operations ij must be fulfilled in the plant r. Therefore:

$$q_{rij}F_{r_{ij}j}(T) = Q_{jr} \tag{198}$$

here:

q_{rij} = the amount of production which can be produced in operation ij per unit of time with plant r.

The sequence of operations established by the process plan for producing tubes (type j) can be expressed in the following manner; not one type of tubes j can be subjected to operation ij until it has passed through the previous operation $(i - 1)j$. This means that the amount of production of type j which has passed through operation ij by moment of time t cannot be greater than the amount of production which has passed through operation $(i - 1)j$ in the same period:

$$q_{rij}F_{r_{ij}j}(t) - q_{r(i-1)j}F_{r_{(i-1)j}j}(t) \leqslant 0 \tag{199}$$

The left hand side of condition (195) can only be equal to zero if the adjusting coefficient (the metal consumption coefficient) is not altered on changing from operation $(i - 1)j$ to ij..

If operation $(i - 1)j$ is accompanied by losses of metal (geometrical) wastage, etc. the following conditions must be fulfilled;

$$\frac{q_{rij}F_{rijj}(t)}{q_{r(i-1)j}F_{r_{(i-1)j}j}(t)} \leqslant \frac{k_{ij}}{K_{(i-1)j}} \tag{200}$$

here:

K_{ij} = the adjusting coefficient (the amount of metal processed to produce a tonne of finished tubes) for operation ij

The problem thus amounts to determining the function $F_{rj}(t)$, with $t > 0$, with which conditions (195)-(200) are satisfied.

The problem can be solved approximately by replacing it by the discrete problem considered in reference (86), in which time intervals t following one another were considered. The shorter the intervals, the more accurately will the problem be solved.

The successive time intervals are given the serial numbers $\tau = 1, 2, \ldots s$. The unit of measurement is the duration of an interval. In this case the quantity q_{rij} expresses the amount of products which will be manufactured in operation ij and plant r in one time interval. We use $X_{\tau rj}$ to denote the time which it is planned to expend during the τ-th time interval on processing type j tubes in equipment r. Compiling the optimum plan amounts to determining values of the variables $X_{\tau rj}$

Obviously the time consumed, during any interval τ, on producing different types of tube j in equipment r cannot exceed the duration of an interval:

$$\sum_{\tau} X_{\tau rj} \leqslant 1 \qquad (201)$$

The total production passing through operation ij in equipment r during different time intervals must be:

$$\sum_{\tau=1}^{s} q_{rij} X_{\tau r_{ij} j} = Q_{ir} \qquad (202)$$

The condition expressing the necessity for observing a prescribed sequence of operations can be written (for ij, where $i>1$) as follows:

$$q_{rij} \sum_{\tau=1}^{\tau_p} X_{\tau r_{ij} j} \leqslant q_{r(i-1)j} \sum_{\tau=1}^{\tau_p} X_{\tau r_{(i-1)j} j} \qquad (203)$$

The change in the adjusting coefficient on transferring from operation $(i-1)j$ to operation ij is expressed by the equation

$$\frac{q_{rij} \sum_{\tau=1}^{s} X_{\tau r_{ij} j}}{q_{r(i-1)j} \sum_{\tau=1}^{s} X_{\tau r_{(i-1)j} j}} \leqslant \frac{K_{ij}}{K_{(i-1)j}} \qquad (204)$$

We will use $\bar{\tau}$ to denote the highest of the numbers τ to which even one value of $X_{\tau rj} > 0$ correspons (consequently $\bar{\tau}$ is the number of the latest of the time intervals during which the operations concerned are carried out). The condition for the optimum plan can be expressed by:

$$\bar{\tau} = min \qquad (205)$$

The treatment of type j tubes in the first equipment r can be subdivided into arbitrary parts carried out in intervals of time not adjacent to one another. The production technology makes it possible to begin operation ij earlier than the entire amount of semi finished products of type j tubes has passed through the preceding operation $(i-1)j$. In this case, however, a certain time may be required, between the end of operation $(i - 1)j$ and the start of operation ij, for moving products and for treatment not taken into account in the mathematical model.

Let ω_{ij} be the shortest permissible time between the start of operation $(i-1)j$ and the start of operation ij, i.e.

$$\omega_{ij} \geqslant \tau_{ij} - \tau_{(i-1)j} \tag{206}$$

Assuming that ω_{ij} and τ_{ij} are whole numbers, the problem can be generalized and approximated to the actual event with the following conditions from reference (83): for ij, where $i>1$

$$\sum_{\tau=1}^{\tau_p} X_{\tau r_{ij}j} \begin{cases} \leqslant \sum_{\tau=1}^{\tau_p-\omega_{ij}} X_{\tau r(i-1)j}, & \text{if } \omega_{ij} < \tau_p \\ = 0, & \text{if } \omega_{ij} \geqslant \tau_p \end{cases} \tag{207}$$

The solution to the problem of compiling a planning schedule is made much more accurate by dividing the total planned amount Q_j of type j tubes into individual batches of $Q'_j < Q_j$. It is necessary, however, to allow for the period of time u_{ij} required for readjusting the plant (changing tools) on changing from one type j of products to another.

* * *

It is possible, by means of mathematical models of technological processes, to solve a more extensive problem; on the basis of the overall production programme for several shops (or works) to determine the optimum range of production for each shop (or works). i.e. to plan the optimum specialization in shops (or works) on a scientific basis, thereby making production as a whole more efficient. The procedure which has been given for calculating optimum process schedules makes it possible to solve an extremely important problem, that of determining the composition of the equipment for shops at the design stage on the basis of the specified volume of production.

Data taken from the standard process charts lie at the basis of the calculation of process schedules using mathematical models. The basic scientific calculation of the standard charts therefore becomes of prime importance. In planning the process schedules which are actually possible it is necessary to establish the optimum parameters of the processes (of extrusion, drawing, cold rolling etc.) for each of the technologies considered:- the optimum length of billet, the dimensions of the blank, the number of tubes per blank, the output, the temperature and velocity conditions, etc. only then commencing the calculation of the standard technological charts, on which is based the mathematical model.

References

1. S I Gubkin: The plastic deformation of metals: Metallurgizdat 1960.
2. S I Gubkin: Theory of the processing of metals by pressure: Metallurgizdat 1947.
3. P S Istomin: The extrusion of metals: Metallurgizdat 1944.
4. I L Perlin: Theory of the extrusion of metals: Metallurgiya 1964.
5. V V Zholobov and G I Zverev: The extrusion of metals: Metallurgizdat 1959.
6. Yu F Shevakin et alia: Tsvetnye Metally 1960, (2), 73-80.
7. E K L Haffer and R M L Elkan: The extrusion of steel tubes and section: Neue Hütte 1960, 5, (5), 266-278.
8. Yu F Shevakin: Groove design and forces in the cold rolling of tubes: Metallurgizdat 1963.
9. Yu F Shevakin et alia: Tsvetnye Metally 1959, (4), 57-63.
10. Yu F Shevakin and A M Rytikov: Tsvetnye Metally 1958, (1), 81-83.
11. Z A Koff: The cold rolling of tubes: Metallurgizdat 1951.
12. Yu F Shevakin and A M Rytikov: Tsvetnye Metally 1958, (12), 70-77.
13. Z A Koff et alia: The cold rolling of tubes: Metallurgizdat 1962.
14. I L Perlin: Drawing theory: Metallurgizdat 1957.
15. L E Al-Shevskii: The tractive forces in the cold drawing of tubes:Metallurgizdat 1952.
16. Yu F Shevakin et alia: Production of non-ferrous tubes: Metallurgizdat 1963.
17. A M Rytikov and Yu F Shevakin: Nauchno-Tekhnicheskii byulleten' TsIIn TsM 1966, 14, (307), 47-49.
18. Block type tube draw benches: Ekspress-informatsiya Prokatka i prokatnoe proizvodstvo 1960, (4), 28-31.
19. Byull.TsIIN TsM 1957, (1), 30.
20. B L Vasil'ev: Hydraulic equipment for extruding non-ferrous metals: Mashgiz 1951.
21. K T Kim et alia: Tsvetnye Metally 1959, (10), 65-68.
22. Sh Gelei: Calculating the forces during the plastic deformation of metals: Metallurgizdat 1956.
23. L V Prozorov: Extrusion of steel sections and tubes: Mashgiz 1951.
24. I Ya Tarnovskii et alia: Theory of the pressure treatment of metals: Metallurgizdat 1963.
25. L V Prozorov: The cold extrusion of thin walled components: Symposium "Recent research into forging technology": Mashgiz 1950.
26. A M Rytikov et alia: The forces on the piercing point during

the extrusion of tubes: Tsvetnye Metally 1964, (2), 70-75.
27. M V Strorozhev and E A Popov: Theory of the pressure treatment of metals: Mashgiz 1957.
28. E Zibel': The treatment of metals in the plastic state: Onti 1934.
29. F I Gofman: Pressing in an enclosed die in a hydraulic press (research into the process): Izdatel'stvo nauchno-tekhnicheskogo obshchestva Moskovskogo Mekhano-Matematicheskogo Instituta 1931.
30. Treco Transactions Amer.Soc.Metals 1962, 55, (4), 697-718.
31. I M Pavlov: Rolling theory: Metallurgizdat 1950.
32. I M Pavlov and V Ya Osadchii: MiSiS Symposium No. XL, Metallurgizdat 1962, p.173-180.
33. Yu F Zil'berg and A P Grundnev: The reasons for alteration in the coefficients of friction during rolling. Rolling theory: Metallurgizdat 1962.
34. V V Zholobov and G I Zverev: Tools for the hot extrusion of metals: Mashinostroenie 1965.
35. Barton, Badsworth and Halling: Revue de Metallurgie 1959, II, p.87.
36. L P Perlin and I Kochish: Symposium of Mintsvetmetzoloto (Ministry of Non-Ferrous Metals and Gold) Works No.28, Metallurgizdat 1958, pp.122-123.
37. N P Besklubenko: Inzhenernyi zhurnal 1961, (2), 193-199.
38. I L Perlin and M B Taubkin: Tsvetnye Metally 1952, (2), 76-81.
39. C Blasly, L Brood, W S Gummer and D B Thompson: J.Inst.Met. 1948, 16.
40. I L Perlin: Tsvetnye Metally 1956, (2), 75-82.
41. Time norms for the principal processes in the cold rolling of tubes: Central Office of Industrial Labour Norms. Moscow 1965.
42. A N Gordov: Fundamentals of pyrometry: "Metallurgiya" 1964.
43. Handbook "Processing of non-ferrous metals and alloys": Metallurgizdat 1961.
44. K V Gagen-Torn: Tsvetnye Metally 1957, (2), 61-67.
45. N S Dobrinskii: Vestnik Mashinostroeniya 1959, (3), 48-52.
46. A M Rytikov et alia: Tsvetnye Metally 1964, (7), 66-68.
47. Yu F Shevakin et alia: Symposium of scientific and technical information concerning the casting and processing of non-ferrous metals and alloys: Issue 6 TsIIN TsM 1965, 33-37.
48. A R E Singer: Metal Industry 1962, 18, 100.
49. R Chadwick: Metal Industry 1962, 12, 100.
50. K Laue: Z für Metallkunde 1960, X, 491.
51. M Z Ermanok and L S Skoblov: Tsvetnye Metally 1963, (10), 78-80.
52. V V Zholobov: Tsvetnye Metally 1959, (8), 62-69.
53. K V Gagen-Torn: Tsvetnye Metally 1957, (9), 82-84.
54. I L Perlin: Symposium of Mintsvetmetzoloto: Works No. 23:

Metallurgizdat 1952, 157-164.
55. I L Perlin: Symposium of Giprotsvetmetobrabotka (Main administration for non-ferrous metal processing works): Issue 21: Metallurgizdat 1963, 183-215.
56. L B Zlotin and Yu P Glebov: Izv VUZ Tsvetnaya Metallurgiya 1962, (4), 169-173.
57. I A Polyakov and P S Tsyrlov: Modern hydraulic tube and rod presses abroad: Issue I: Metallurgizdat 1957.
58. G B Dantzig: Maximisation of linear function of variables subject to linear inequalities: "Activity analysis of production and allocation": 1951, (4).
59. A M Rytikov and Yu F Shevakin: Tsvetnye Metally 1966, (5), 89-93.
60. A P Smiryagin: Industrial non-ferrous metals and alloys: Metallurgizdat 1956.
61. M K Rybal'Chenko and A A Rastorchuk: Properties of metals and alloys (Handbook): Metallurgizdat 1949.
62. A V Tret'yakov et alia: Mechanical properties of metals and alloys when pressure treated: Metallurgiya 1964.
63. K N Bogoyavlenskii et alia: Pressure treatment of non-ferrous metals and alloys: Metallurgizdat 1964.
64. M M Bernshtein: Trudy UkrNITI Issue 1, Metallurgizdat 1959, 125-144.
65. V Ya Shapiro: Tsvetnye Metally 1959, (9), 68-74.
66. K V Gagen-Torn and S V Veretennikov: Symposium "Tubes drawn with a floating mandrel": TsBTI Sverdlovsk 1959, 64-69.
67. S M Egorov and A M Komarov: Imperial Chemical Industries Works for producing non-ferrous metal tubes: Byull.TsIIN TsM 1957, 18, (95), 37-41.
68. B A Egorov and K T Kim: Tsvetnye Metally 1962, (9), 79-85.
69. L M Radchenko: Byull. Tsvetnye Metallurgiya 1962, (7), 47-49.
70. I M Razumov and E G Ginzburg: Organisation of production at works processing non-ferrous metals: Metallurgizdat 1956.
71. I L Perlin and L A Grinberg: Symposium of Mintsvetmetzoloto scientific works: Issue XX Metallurgizdat 1950, 269.
72. N P Belousov: Works of the Inter-University Scientific and Technical Conference on "Modern Developments in Rolling Mill Production": Vol.I Leningrad Polytechnic Institute 1958, 234-239.
73. V A Kochkin: Metallurg. 1937, (5), 39-42.
74. N A Shaposhnikov: Mechanical tests on metals: Mashgiz 1951.
75. I N Bronshtein and K A Semendyaev: Handbook on mathematics: Gostekhteoretizdat 1957.
76. S I Borisov and M M Bernshtein: VNITI Bulletin of Scientific and Technical Information No.4: Metallurgizdat 1958.
77. Yu F Shevakin and A M Rytikov: Tsvetnye Metally 1958, (12), 70-77.

78. Yu F Shevakin and A M Rytikov: Tsvetnye Metally 1961, (4), 51-58.
79. A I Tselikov: Fundamentals of rolling theory: Metallurgiya 1965.
80. Symposium "The pressure treatment of metals": Works of the Institute of Steel and Alloys and the Moscow Power Engineering Institute: Issue 61, 1965, II, 9-19, 239-248.
81. P K Teterin: Stal' 1953, (12), 1106-1111.
82. Yu F Shevakin and F S Seidaliev: Tube cold rolling mills: Metallurgiya 1965.
83. A L Lur'e: Mathematical methods of solving problems to find the output when planning Socialist economy: Nauka 1964.
84. M Z Ermanok: Tsvetnye Metally 1960, (4), 79-82.
85. Iron and Steel Engineer: Pittsburg Penn.(USA) 1957, (5).
86. Steel 1964, 194, (8), 134-141.
87. Pavlov et alia: Tsvetnye Metally 1955, (6), 41-50.
88. N P Belousov et alia: Leningrad Polytechnic Institute Works No.243: Mashinostroenie 1965, 141-145.
89. Yu F Shevakin et alia: Nauchno-tekhnicheskii Byulleten' TsIIN TsM 1965, 4, (273), 41-45.
90. V A Plokhov: Tsvetnye Metally 1959, (12), 67-69.
91. V V Zholobov: Tsvetnye Metally 1960, (4), 76-79.
92. K B Gagen-Torn: TsIIN TsM Symposium 1962, 6-21.
93. N I Kovalev: Computer technique in planning: "Ekonomika" 1964.
94. L V Kantorovich and V K Zalgaller: Calculating the rational cutting up of industrial materials: Lenizdat 1951.
95. M Z Ermanok et alia: Tekhnologiya legkikh splavov 1963, (1), 47-51.